Pitman Research Notes in Mathematics Series

Submission of proposals for consideration

Suggestions for publication, in the form of outlines and representative samples, are invited by the Editorial Board for assessment. Intending authors should approach one of the main editors or another member of the Editorial Board, citing the relevant AMS subject classifications. Alternatively, outlines may be sent directly to the publisher's offices. Refereeing is by members of the board and other mathematical authorities in the topic concerned, throughout the world.

Preparation of accepted manuscripts

On acceptance of a proposal, the publisher will supply full instructions for the preparation of manuscripts in a form suitable for direct photo-lithographic reproduction. Specially printed grid sheets can be provided and a contribution is offered by the publisher towards the cost of typing. Word processor output, subject to the publisher's approval, is also acceptable.

Illustrations should be prepared by the authors, ready for direct reproduction without further improvement. The use of hand-drawn symbols should be avoided wherever possible, in order to maintain maximum clarity of the text.

The publisher will be pleased to give any guidance necessary during the preparation of a typescript, and will be happy to answer any queries.

Important note

In order to avoid later retyping, intending authors are strongly urged not to begin final preparation of a typescript before receiving the publisher's guidelines. In this way it is hoped to preserve the uniform appearance of the series.

Longman Scientific & Technical
Longman House
Burnt Mill
Harlow, Essex, CM20 2JE
UK
(Telephone (0279) 426721)

Titles in this series. A full list is available on request from the publisher.

51 Subnormal operators
J B Conway
52 Wave propagation in viscoelastic media
F Mainardi
53 Nonlinear partial differential equations and their applications: Collège de France Seminar. Volume I
H Brezis and J L Lions
54 Geometry of Coxeter groups
H Hiller
55 Cusps of Gauss mappings
T Banchoff, T Gaffney and C McCrory
56 An approach to algebraic K-theory
A J Berrick
57 Convex analysis and optimization
J-P Aubin and R B Vintner
58 Convex analysis with applications in the differentiation of convex functions
J R Giles
59 Weak and variational methods for moving boundary problems
C M Elliott and J R Ockendon
60 Nonlinear partial differential equations and their applications: Collège de France Seminar. Volume II
H Brezis and J L Lions
61 Singular Systems of differential equations II
S L Campbell
62 Rates of convergence in the central limit theorem
Peter Hall
63 Solution of differential equations by means of one-parameter groups
J M Hill
64 Hankel operators on Hilbert Space
S C Power
65 Schrödinger-type operators with continuous spectra
M S P Eastham and H Kalf
66 Recent applications of generalized inverses
S L Campbell
67 Riesz and Fredholm theory in Banach algebra
B A Barnes, G J Murphy, M R F Smyth and T T West
68 Evolution equations and their applications
K Kappel and W Schappacher
69 Generalized solutions of Hamilton–Jacobi equations
P L Lions
70 Nonlinear partial differential equations and their applications: Collège de France Seminar. Volume III
H Brezis and J L Lions
71 Spectral theory and wave operators for the Schrödinger equation
A M Berthier
72 Approximation of Hilbert space operators I
D A Herrero
73 Vector valued Nevanlinna theory
H J W Ziegler
74 Instability, nonexistence and weighted energy methods in fluid dynamics and related theories
B Straughan
75 Local bifurcation and symmetry
A Vanderbauwhede
76 Clifford analysis
F Brackx, R Delanghe and F Sommen
77 Nonlinear equivalence, reduction of PDEs to ODEs and fast convergent numerical methods
E E Rosinger
78 Free boundary problems, theory and applications. Volume I
A Fasano and M Primicerio
79 Free boundary problems, theory and applications. Volume II
A Fasano and M Primicerio
80 Symplectic geometry
A Crumeyrolle and J Grifone
81 An algorithmic analysis of a communication model with retransmission of flawed messages
D M Lucantoni
82 Geometric games and their applications
W H Ruckle
83 Additive groups of rings
S Feigelstock
84 Nonlinear partial differential equations and their applications: Collège de France Seminar. Volume IV
H Brezis and J L Lions
85 Multiplicative functionals on topological algebras
T Husain
86 Hamilton–Jacobi equations in Hilbert spaces
V Barbu and G Da Prato
87 Harmonic maps with symmetry, harmonic morphisms and deformations of metric
P Baird
88 Similarity solutions of nonlinear partial differential equations
L Dresner
89 Contributions to nonlinear partial differential equations
C Bardos, A Damlamian, J I Díaz and J Hernández
90 Banach and Hilbert spaces of vector-valued functions
J Burbea and P Masani
91 Control and observation of neutral systems
D Salamon
92 Banach bundles, Banach modules and automorphisms of C^*-algebras
M J Dupré and R M Gillette
93 Nonlinear partial differential equations and their applications: Collège de France Seminar. Volume V
H Brezis and J L Lions
94 Computer algebra in applied mathematics: an introduction to MACSYMA
R H Rand
95 Advances in nonlinear waves. Volume I
L Debnath
96 FC-groups
M J Tomkinson
97 Topics in relaxation and ellipsoidal methods
M Akgül
98 Analogue of the group algebra for topological semigroups
H Dzinotyiweyi
99 Stochastic functional differential equations
S E A Mohammed

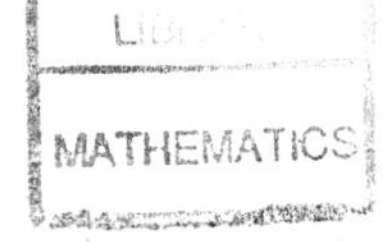

100 Optimal control of variational inequalities
V Barbu
101 Partial differential equations and dynamical systems
W E Fitzgibbon III
102 Approximation of Hilbert space operators Volume II
C Apostol, L A Fialkow, D A Herrero and D Voiculescu
103 Nondiscrete induction and iterative processes
V Ptak and F-A Potra
104 Analytic functions – growth aspects
O P Juneja and G P Kapoor
105 Theory of Tikhonov regularization for Fredholm equations of the first kind
C W Groetsch
106 Nonlinear partial differential equations and free boundaries. Volume I
J I Díaz
107 Tight and taut immersions of manifolds
T E Cecil and P J Ryan
108 A layering method for viscous, incompressible L_p flows occupying R^n
A Douglis and E B Fabes
109 Nonlinear partial differential equations and their applications: Collège de France Seminar. Volume VI
H Brezis and J L Lions
110 Finite generalized quadrangles
S E Payne and J A Thas
111 Advances in nonlinear waves. Volume II
L Debnath
112 Topics in several complex variables
E Ramírez de Arellano and D Sundararaman
113 Differential equations, flow invariance and applications
N H Pavel
114 Geometrical combinatorics
F C Holroyd and R J Wilson
115 Generators of strongly continuous semigroups
J A van Casteren
116 Growth of algebras and Gelfand–Kirillov dimension
G R Krause and T H Lenagan
117 Theory of bases and cones
P K Kamthan and M Gupta
118 Linear groups and permutations
A R Camina and E A Whelan
119 General Wiener–Hopf factorization methods
F-O Speck
120 Free boundary problems: applications and theory. Volume III
A Bossavit, A Damlamian and M Fremond
121 Free boundary problems: applications and theory. Volume IV
A Bossavit, A Damlamian and M Fremond
122 Nonlinear partial differential equations and their applications: Collège de France Seminar. Volume VII
H Brezis and J L Lions
123 Geometric methods in operator algebras
H Araki and E G Effros
124 Infinite dimensional analysis–stochastic processes
S Albeverio
125 Ennio de Giorgi Colloquium
P Krée
126 Almost-periodic functions in abstract spaces
S Zaidman
127 Nonlinear variational problems
A Marino, L Modica, S Spagnolo and M Degliovanni
128 Second-order systems of partial differential equations in the plane
L K Hua, W Lin and C-Q Wu
129 Asymptotics of high-order ordinary differential equations
R B Paris and A D Wood
130 Stochastic differential equations
R Wu
131 Differential geometry
L A Cordero
132 Nonlinear differential equations
J K Hale and P Martinez-Amores
133 Approximation theory and applications
S P Singh
134 Near-rings and their links with groups
J D P Meldrum
135 Estimating eigenvalues with *a posteriori/a priori* inequalities
J R Kuttler and V G Sigillito
136 Regular semigroups as extensions
F J Pastijn and M Petrich
137 Representations of rank one Lie groups
D H Collingwood
138 Fractional calculus
G F Roach and A C McBride
139 Hamilton's principle in continuum mechanics
A Bedford
140 Numerical analysis
D F Griffiths and G A Watson
141 Semigroups, theory and applications. Volume I
H Brezis, M G Crandall and F Kappel
142 Distribution theorems of L-functions
D Joyner
143 Recent developments in structured continua
D De Kee and P Kaloni
144 Functional analysis and two-point differential operators
J Locker
145 Numerical methods for partial differential equations
S I Hariharan and T H Moulden
146 Completely bounded maps and dilations
V I Paulsen
147 Harmonic analysis on the Heisenberg nilpotent Lie group
W Schempp
148 Contributions to modern calculus of variations
L Cesari
149 Nonlinear parabolic equations: qualitative properties of solutions
L Boccardo and A Tesei
150 From local times to global geometry, control and physics
K D Elworthy

151 A stochastic maximum principle for optimal control of diffusions
U G Haussmann
152 Semigroups, theory and applications. Volume II
H Brezis, M G Crandall and F Kappel
153 A general theory of integration in function spaces
P Muldowney
154 Oakland Conference on partial differential equations and applied mathematics
L R Bragg and J W Dettman
155 Contributions to nonlinear partial differential equations. Volume II
J I Díaz and P L Lions
156 Semigroups of linear operators: an introduction
A C McBride
157 Ordinary and partial differential equations
B D Sleeman and R J Jarvis
158 Hyperbolic equations
F Colombini and M K V Murthy
159 Linear topologies on a ring: an overview
J S Golan
160 Dynamical systems and bifurcation theory
M I Camacho, M J Pacifico and F Takens
161 Branched coverings and algebraic functions
M Namba
162 Perturbation bounds for matrix eigenvalues
R Bhatia
163 Defect minimization in operator equations: theory and applications
R Reemtsen
164 Multidimensional Brownian excursions and potential theory
K Burdzy
165 Viscosity solutions and optimal control
R J Elliott
166 Nonlinear partial differential equations and their applications: Collège de France Seminar. Volume VIII
H Brezis and J L Lions
167 Theory and applications of inverse problems
H Haario
168 Energy stability and convection
G P Galdi and B Straughan
169 Additive groups of rings. Volume II
S Feigelstock
170 Numerical analysis 1987
D F Griffiths and G A Watson
171 Surveys of some recent results in operator theory. Volume I
J B Conway and B B Morrel
172 Amenable Banach algebras
J-P Pier
173 Pseudo-orbits of contact forms
A Bahri
174 Poisson algebras and Poisson manifolds
K H Bhaskara and K Viswanath
175 Maximum principles and eigenvalue problems in partial differential equations
P W Schaefer
176 Mathematical analysis of nonlinear, dynamic processes
K U Grusa
177 Cordes' two-parameter spectral representation theory
D F McGhee and R H Picard
178 Equivariant K-theory for proper actions
N C Phillips
179 Elliptic operators, topology and asymptotic methods
J Roe
180 Nonlinear evolution equations
J K Engelbrecht, V E Fridman and E N Pelinovski
181 Nonlinear partial differential equations and their applications: Collège de France Seminar. Volume IX
H Brezis and J L Lions
182 Critical points at infinity in some variational problems
A Bahri
183 Recent developments in hyperbolic equations
L Cattabriga, F Colombini, M K V Murthy and S Spagnolo
184 Optimization and identification of systems governed by evolution equations on Banach space
N U Ahmed
185 Free boundary problems: theory and applications. Volume I
K H Hoffmann and J Sprekels
186 Free boundary problems: theory and applications. Volume II
K H Hoffmann and J Sprekels
187 An introduction to intersection homology theory
F Kirwan
188 Derivatives, nuclei and dimensions on the frame of torsion theories
J S Golan and H Simmons
189 Theory of reproducing kernels and its applications
S Saitoh
190 Volterra integrodifferential equations in Banach spaces and applications
G Da Prato and M Iannelli
191 Nest algebras
K R Davidson
192 Surveys of some recent results in operator theory. Volume II
J B Conway and B B Morrel
193 Nonlinear variational problems. Volume II
A Marino and M K V Murthy
194 Stochastic processes with multidimensional parameter
M E Dozzi
195 Prestressed bodies
D Iesan
196 Hilbert space approach to some classical transforms
R H Picard
197 Stochastic calculus in application
J R Norris
198 Radical theory
B J Gardner
199 The C^*-algebras of a class of solvable Lie groups
X Wang
200 Stochastic analysis, path integration and dynamics
K D Elworthy and J C Zambrini

201 Riemannian geometry and holonomy groups
S Salamon
202 Strong asymptotics for extremal errors and polynomials associated with Erdős type weights
D S Lubinsky
203 Optimal control of diffusion processes
V S Borkar
204 Rings, modules and radicals
B J Gardner
205 Two-parameter eigenvalue problems in ordinary differential equations
M Faierman
206 Distributions and analytic functions
R D Carmichael and D Mitrovic
207 Semicontinuity, relaxation and integral representation in the calculus of variations
G Buttazzo
208 Recent advances in nonlinear elliptic and parabolic problems
P Bénilan, M Chipot, L Evans and M Pierre
209 Model completions, ring representations and the topology of the Pierce sheaf
A Carson
210 Retarded dynamical systems
G Stepan
211 Function spaces, differential operators and nonlinear analysis
L Paivarinta
212 Analytic function theory of one complex variable
C C Yang, Y Komatu and K Niino
213 Elements of stability of visco-elastic fluids
J Dunwoody
214 Jordan decomposition of generalized vector measures
K D Schmidt
215 A mathematical analysis of bending of plates with transverse shear deformation
C Constanda
216 Ordinary and partial differential equations. Volume II
B D Sleeman and R J Jarvis
217 Hilbert modules over function algebras
R G Douglas and V I Paulsen
218 Graph colourings
R Wilson and R Nelson
219 Hardy-type inequalities
A Kufner and B Opic
220 Nonlinear partial differential equations and their applications: Collège de France Seminar. Volume X
H Brezis and J L Lions
221 Workshop on dynamical systems
E Shiels and Z Coelho
222 Geometry and analysis in nonlinear dynamics
H W Broer and F Takens
223 Fluid dynamical aspects of combustion theory
M Onofri and A Tesei
224 Approximation of Hilbert space operators. Volume I. 2nd edition
D Herrero
225 Operator theory: proceedings of the 1988 GPOTS–Wabash conference
J B Conway and B B Morrel
226 Local cohomology and localization
J L Bueso Montero, B Torrecillas Jover and A Verschoren
227 Nonlinear waves and dissipative effects
D Fusco and A Jeffrey
228 Numerical analysis 1989
D F Griffiths and G A Watson
229 Recent developments in structured continua. Volume III
D De Kee and P Kaloni
230 Boolean methods in interpolation and approximation
F J Delvos and W Schempp
231 Further advances in twistor theory. Volume I
L J Mason and L P Hughston
232 Further advances in twistor theory. Volume II
L J Mason and L P Hughston
233 Geometry in the neighborhood of invariant manifolds of maps and flows and linearization
U Kirchgraber and K Palmer
234 Quantales and their applications
K I Rosenthal
235 Integral equations and inverse problems
V Petkov and R Lazarov
236 Pseudo-differential operators
S R Simanca
237 A functional analytic approach to statistical experiments
I M Bomze
238 Quantum mechanics, algebras and distributions
D Dubin and M Hennings
239 Hamilton flows and evolution semigroups
J Gzyl
240 Topics in controlled Markov chains
V S Borkar
241 Invariant manifold theory for hydrodynamic transition
S Sritharan
242 Lectures on the spectrum of $L^2(\Gamma\backslash G)$
F L Williams
243 Progress in variational methods in Hamiltonian systems and elliptic equations
M Girardi, M Matzeu and F Pacella
244 Optimization and nonlinear analysis
A Ioffe, M Marcus and S Reich
245 Inverse problems and imaging
G F Roach
246 Semigroup theory with applications to systems and control
N U Ahmed
247 Periodic-parabolic boundary value problems and positivity
P Hess
248 Distributions and pseudo-differential operators
S Zaidman
249 Progress in partial differential equations: the Metz surveys
M Chipot and J Saint Jean Paulin
250 Differential equations and control theory
V Barbu

251 Stability of stochastic differential equations with respect to semimartingales
X Mao
252 Fixed point theory and applications
J Baillon and M Théra
253 Nonlinear hyperbolic equations and field theory
M K V Murthy and S Spagnolo
254 Ordinary and partial differential equations. Volume III
B D Sleeman and R J Jarvis
255 Harmonic maps into homogeneous spaces
M Black
256 Boundary value and initial value problems in complex analysis: studies in complex analysis and its applications to PDEs 1
R Kühnau and W Tutschke
257 Geometric function theory and applications of complex analysis in mechanics: studies in complex analysis and its applications to PDEs 2
R Kühnau and W Tutschke
258 The development of statistics: recent contributions from China
X R Chen, K T Fang and C C Yang
259 Multiplication of distributions and applications to partial differential equations
M Oberguggenberger
260 Numerical analysis 1991
D F Griffiths and G A Watson
261 Schur's algorithm and several applications
M Bakonyi and T Constantinescu

M Bakonyi

The College of William and Mary, Williamsburg, USA

and

T Constantinescu

Stanford University, USA

Schur's algorithm and several applications

Copublished in the United States with
John Wiley & Sons, Inc., New York

Longman Scientific & Technical
Longman Group UK Limited
Longman House, Burnt Mill, Harlow
Essex CM20 2JE, England
and Associated companies throughout the world.

Copublished in the United States with
John Wiley & Sons Inc., 605 Third Avenue, New York, NY 10158

First published 1992

AMS Subject Classification: 47A20, 47A56, 15A57, 42C05, 93

ISSN 0269-3674

ISBN 0 582 09120 9

British Library Cataloguing in Publication Data

A catalogue record for this book is
available from the British Library

Library of Congress Cataloging-in-Publication Data

Bakonyi, M. (Mihály), 1962–
Schur's algorithm and several applications / M. Bakonyi and T. Constantinescu.
p. cm. -- (Pitman research notes in mathematics series ISSN 0269-3674 ; 261)
Includes bibliographical references and index.
1. Dilation theory (Operator theory) 2. Analytic functions.
I. Constantinescu, T. (Tiberiu), 1955– . II. Title. III. Series.
QA329.B35 1992
515'.724--dc20 92-4758
CIP

Printed and bound in Great Britain
by Biddles Ltd, Guildford and King's Lynn

Contents

Preface

The algorithm of I. Schur is a remarkable example of an elementary result with applications that cover a large area and has deep implications in fields among which we mention operatorial matrix completions, interpolation and dilation theory, orthogonal polynomials and electrical engineering.

The aim of this book is to present as many results and applications of the ideas in Schur's 'Über Potenzreihen, die im Innern die Einheitkreises beschränkt sind' as possible. We stick to the initial background of the theory of analytic functions in the unit disc, so that this book can be also regarded as an introduction to other volumes we intend to write in connection with Schur's algorithm. Thus non–stationary developments, connections with indefinite inner product spaces as well as continuous analogues of the problems discussed in this volume are to be described in these future volumes.

Other methods of approach have been developed for many of the problems we are faced with. First, we mention the geometric approach based on the Sz. Nagy-Foiaş commutant lifting theorem to interpolation problems. A comprehensive description of this approach with applications in geophysics, control theory and signal processing can be found in the recent book by Ciprian Foiaş and Arthur Frazho, *The Commutant Lifting Approach to Interpolation Problems*, Birkhäuser Verlag, Basel, 1990. The realisation theory of matrix or operator–valued functions originated in system theory, where the functions appear as transfer functions and their realisation provides many insights into the behaviour of the systems. Using this view, many interpolation problems for rational matrix functions have been studied in the book by Joseph Ball, Israel Gohberg and Leiba Rodman, *Interpolation for Rational Matrix Functions*, Birkhäuser Verlag, Basel, 1990. We also mention the reproducing kernel Hilbert space approach to interpolation problems, which originates in a paper of Béla Sz. Nagy and Ádám Korányi. Significant contributions over the years are due to Louis de Branges, Marvin Rosenblum and James Rovnyak (see [74]). For this method, the recent and new developments of the problems are contained in the monograph by Harry Dym (see [40]). The shift–invariant subspace approach by Joseph Ball and Joseph Helton also produced significant results in this subject (see [12]). Their key tool is the celebrated Beurling–Lax theorem. An approach that extracted the important algebraic properties of the various extension problems was initiated by Harry Dym and Israel Gohberg, who also recognised the importance of maximum entropy. The band method, was further developed by Israel Gohberg, Marinus Kaashoek and Hugo Woerdeman (*J. Operator Theory* **22** (1989), 109–155). Another approach to interpolation and fast factorisation of structured matrices based on conservation of energy considerations (using the Lyapunov equation) was developed by Thomas Kailath and his school.

Our goal is to present the basic results of interpolation theory from the unitary point of view of Schur analysis and to extend this method of investigation to new domains such as spectral factorisations and orthogonal polynomials. In our last chapter we connect the methods of Schur analysis with a method using extensions of isometries to unitaries, first developed by M. A. Naimark. In this way we solve several interpolation problems and treat the lifting of commutants.

Most of the material was prepared while the authors were in the Department of Mathematics at INCREST, Bucharest. Several of the problems have also been presented at the University of Timişoara and we are grateful for the fact that Professor Dumitru Gaşpar made possible the publishing of a preliminary version of these notes.

The idea of publishing the revised version of this book in the Pitman Research Notes in Mathematics series was suggested to us by Professor Ciprian Foiaş and Professor Israel Gohberg for which we are grateful. We thank Professor Ronald Douglas, editor of this series for helping us to publish the present version.

The authors would like to thank their present institutions, The College of William and Mary, and Standford University for support while preparing the final version of these notes. We thank Professor Thomas Kailath for his support, interest and guidance during the past years. We are also grateful for the support of Professor Charles Johnson. Finally, we thank all contributors who helped us with suggestions and comments.

We also thank our wives, Gina Bakonyi and Maria Constantinescu, for their patience and understanding, and we hope that their forbearance will continue.

Mihály Bakonyi
Williamsburg, Virginia

Tiberiu Constantinescu
Stanford, California

September, 1991

CHAPTER 1

Schur algorithms

Initially viewed as a continued fraction algorithm, Schur's algorithm turns out to be the precursor of the layer–peeling idea used in a wide class of applications drawn out from the transmission line model, which is its visual counterpart. In this chapter we treat in some detail the classical Schur algorithm together with an operatorial variant. Applications to classical extrapolation problems – the first motivation of Schur's work – are indicated. We conclude with an account of discrete transmission line models.

1.1 The scalar Schur algorithm

We denote by H^∞ the Hardy space of bounded analytic functions on the unit disc $\mathbf{D} = \{z \in \mathbf{C} \mid\mid z \mid< 1\}$ and by $\mathcal{S}$ the unit ball of H^∞, $\mathcal{S} = \{f \in H^\infty \mid\mid\mid F \mid\mid_\infty = \sup_{z\in\mathbf{D}} \mid F(z) \mid\leq 1\}$. The set $\mathcal{S}$ is sometimes called the Schur class of analytic functions. The conformal maps of $\mathbf{D}$ onto $\mathbf{D}$ can be explicitly written as $\lambda\frac{z-\alpha}{1-\bar{\alpha}z}$ with $\mid \lambda \mid= 1$ and $\alpha \in \mathbf{D}$, and as a consequence of this fact and of Schwartz's Lemma, taking $F \in \mathcal{S}$, $\gamma = F(0)$ and

$$F_1(z) = \frac{F(z) - \gamma}{z(1 - \bar{\gamma}F(z))}$$

we have that $F_1 \in \mathcal{S}$. This remark leads to the following algorithm of I. Schur

Schur's algorithm. Fix $F \in \mathcal{S}$ and define

$$F_o = F, \quad \gamma_n = F_n(0), \quad n \geq 0$$
$$F_{n+1} = \frac{F_n(z) - \gamma_n}{z(1 - \bar{\gamma}_n F_n(z))}.$$

The parameters $\{\gamma_n\}_{n\geq 0}$ in the above algorithm are called the Schur parameters of the function F. Before stating the main result concerning the Schur algorithm a piece of notation is also necessary: for a polynomial P of degree n,$P^T(z) = z^n\bar{P}(1/z)$ and $\bar{P}(z) = \overline{P(\bar{z})}$.

Theorem 1.1 The Schur algorithm realises a one–to–one correspondence between the Schur class $\mathcal{S}$ and the set of sequences of complex numbers $\{\gamma_n\}_{n\geq 0}$ having the properties:$\mid \gamma_n \mid\leq 1$ for $n \geq 0$ and when, for a certain $n_o \in \mathbf{N}$, $\mid \gamma_{n_o} \mid= 1$ then $\gamma_n = 0$ for $n > n_o$. The situation when there exists an $n_o \in \mathbf{N}$ such that $\mid \gamma_{n_o} \mid= 1$ appears exactly when the function for which we apply the Schur algorithm is a finite Blaschke product of degree n_o.

Proof Let $F \in \mathcal{S}$. As $F_n \in \mathcal{S}$, we find that $\mid \gamma_n \mid\leq 1$. By virtue of the maximum principle, if $\mid \gamma_{n_o} \mid= 1$ then $F_{n_o}(z) = \gamma_{n_o}$, $z \in \mathbf{D}$ and $F_n \equiv 0$ for $n > n_o$, $\gamma_n = 0$ for $n > n_o$. This proves that the Schur parameters have the required properties. Then, we remark that from the Schur algorithm,

$$F_n(z) = \frac{\gamma_n + zF_{n+1}(z)}{1 + z\bar{\gamma}_n F_{n+1}(z)} \tag{1.1}$$

and writing a finite Blaschke product of degree m as $\lambda P^T(z)/P(z)$ with $\mid \lambda \mid= 1$ and P a polynomial of degree m, we deduce from (1.1) that F_{n+1} is a Blaschke product of degree m if and only if F_n is a Blaschke produce of degree $m+1$.

Consequently, we have an $n_o \in \mathbf{N}$ with $\mid \gamma_{n_o} \mid= 1$ if and only if F is a finite Blaschke product of degree n_o.

We now show that the Schur parameters uniquely determine the function F, and for this purpose we have to consider only the case when $\mid \gamma_n \mid< 1$ for $n \geq 0$.

By (1.1) we deduce that:

$$F(z) = \frac{\mathcal{A}_n(z) + z\mathcal{B}_n^T(z)F_{n+1}(z)}{\mathcal{B}_n(z) + z\mathcal{A}_n^T(z)F_{n+1}(z)} \tag{1.2}$$

where $\mathcal{A}_n$ and $\mathcal{B}_n$ are polynomials of degree n, usually called the Schur polynomials. These polynomials satisfy the following recurrence relations:

$$\mathcal{A}_o(z) = \gamma_o, \qquad \mathcal{B}_o(z) = 1,$$

and for $n \geq 0$,

$$\mathcal{A}_{n+1}(z) = \mathcal{A}_n(z) + z\gamma_{n+1}\mathcal{B}_n^T(z) \tag{1.3}$$

$$\mathcal{B}_{n+1}(z) = \mathcal{B}_n(z) + z\gamma_{n+1}\mathcal{A}_n^T(z). \tag{1.4}$$

We can use a matricial transcription of these relations,

$$\begin{bmatrix} \mathcal{B}_{n+1}(z) & \mathcal{A}_{n+1}(z) \\ \mathcal{A}^T_{n+1}(z) & \mathcal{B}^T_{n+1}(z) \end{bmatrix} = \begin{bmatrix} 1 & z\gamma_{n+1} \\ \bar{\gamma}_{n+1} & z \end{bmatrix} \begin{bmatrix} \mathcal{B}_n(z) & \mathcal{A}_n(z) \\ \mathcal{A}^T_n(z) & \mathcal{B}^T_n(z) \end{bmatrix}. \tag{1.5}$$

We remark that

$$\begin{bmatrix} 1 & z\gamma_n \\ \bar{\gamma}_n & z \end{bmatrix} \begin{bmatrix} -1 & 0 \\ 0 & 1 \end{bmatrix} \begin{bmatrix} 1 & z\gamma_n \\ \bar{\gamma}_n & z \end{bmatrix}^* \leq (1- \mid \gamma_n \mid^2) \begin{bmatrix} -1 & 0 \\ 0 & 1 \end{bmatrix}$$

with equality for $\mid z \mid = 1$, and using (1.5) it follows that:

$$\begin{bmatrix} \mathcal{B}_n(z) & \mathcal{A}_n(z) \\ \mathcal{A}^T_n(z) & \mathcal{B}^T_n(z) \end{bmatrix} \begin{bmatrix} -1 & 0 \\ 0 & 1 \end{bmatrix} \begin{bmatrix} \mathcal{B}_n(z) & \mathcal{A}_n(z) \\ \mathcal{A}^T_n(z) & \mathcal{B}^T_n(z) \end{bmatrix}^* \leq \prod_{k=0}^{n} (1- \mid \gamma_k \mid^2) \begin{bmatrix} -1 & 0 \\ 0 & 1 \end{bmatrix} \tag{1.6}$$

with equality for $\mid z \mid = 1$. The first consequence of (1.6) is that for $z \in \bar{\mathbf{D}}$,

$$\mid \mathcal{B}_n(z) \mid^2 - \mid \mathcal{A}_n(z) \mid^2 \geq \prod_{k=0}^{n} (1- \mid \gamma_k \mid^2) \tag{1.7}$$

and so $\mathcal{B}_n$ has no zeros in $\bar{\mathbf{D}}$.

By taking determinants on the both sides of (1.5), we also derive that for $z \in \bar{\mathbf{D}}$,

$$\mathcal{B}_n(z)\mathcal{B}^T_n(z) - \mathcal{A}_n(z)\mathcal{A}^T_n(z) = z^n \prod_{k=0}^{n} (1- \mid \gamma_k \mid^2) \qquad (1.8)P$$

Finally, using (1.2) and (1.8), we obtain:

$$F(z) - \frac{\mathcal{A}_n(z)}{\mathcal{B}_n(z)} = z^{n+1} \prod_{k=0}^{n} (1- \mid \gamma_k \mid^2) \frac{F_{n+1}(z)}{\mathcal{B}_n(z)(\mathcal{B}_n(z) + z\mathcal{A}^T_n(z)F_{n+1}(z))}$$

which shows that the sequence $\mathcal{A}_n/\mathcal{B}_n$, $n \geq 0$ converges uniformly on compact subsets of $\mathbf{D}$ to F. But $\mathcal{A}_n$ and $\mathcal{B}_n$ are uniquely determined by $\{\gamma_k\}_{k=0}^n$ and so F will be uniquely determined by its Schur parameters. ∎

Remark 1.2 When the Schur algorithm works for a function $F \in \mathcal{S}$ some other functions in $\mathcal{S}$ are naturally imposed. Thus, there appear the functions F_n and also the functions $\mathcal{A}_n/\mathcal{B}_n$ which are in $\mathcal{S}$ by (1.7). Being given the Schur parameters of F, we can also obtain the Schur parameters of these functions. First, $F_n(0) = \gamma_n$, and applying the Schur algorithm to F_n,

$$(F_n)_1(z) = \frac{F_n(z) - \gamma_n}{z(1 - \bar{\gamma}_n F_n(z))};$$

by (1.1), it turns out that

$$(F_n)_1 = F_{n+1}.$$

Further, for $k \geq 0$, we have by induction

$$(F_n)_k = F_{n+k}$$

and this shows that the Schur parameters of F_n are $\{\gamma_{n+k}\}_{k\geq 0}$.

For determining the Schur parameters of $\mathcal{A}_n/\mathcal{B}_n$, we define the function $F^{(n)}$ associated to the Schur parameters $\{\gamma_o, \gamma_1, ..., \gamma_n, 0, 0, ...\}$. By the proof of Theorem 1.1 (and with a self–evident notation),

$$\begin{aligned} \mathcal{A}_k(F^{(n)}) &= \mathcal{A}_k(\{\gamma_m\}_{m=0}^n) \\ \mathcal{B}_k(F^{(n)}) &= \mathcal{B}_k(\{\gamma_m\}_{m=0}^n), \quad 0 \leq k \leq n. \end{aligned}$$

As $(F^{(n)})_{n+1} \equiv 0$, we use (1.2) in order to obtain

$$F^{(n)} = \frac{\mathcal{A}_n(\{\gamma_m\}_{m=0}^n)}{\mathcal{B}_n(\{\gamma_m\}_{m=0}^n)}$$

which means that the Schur parameters of $\mathcal{A}_n/\mathcal{B}_n$ are $\{\gamma_o, \gamma_1, ..., \gamma_n, 0, 0, ...\}$.

In the proof of Theorem 1.1 we remarked that the Schur algorithm plays a role also in the rational approximation of the functions in the class $\mathcal{S}$. A more precise result can now be stated in the following theorem.

Theorem 1.3 For $F \in \mathcal{S}$ there exists a sequence of finite Blaschke products converging to F uniformly on compact subsets of $\mathbf{D}$.

Proof For $F \in \mathcal{S}$, we take $\{\gamma_n\}_{n\geq 0}$ to be its Schur parameters and by virtue of Theorem 1.1, we can suppose that $| \gamma_n |< 1$, for all $n \geq 0$. For $n \geq 0$ we define (also based on Theorem 1.1) the finite Blaschke product B_n given by the Schur parameters

$\{\gamma_o, \gamma_1, \cdots, \gamma_n, 1\}$. By (1.2),

$$B_n(z) = \frac{\mathcal{A}_n(z) + z\mathcal{B}_n^T(z)}{\mathcal{B}_n(z) + z\mathcal{A}_n^T(z)}$$

and also using (1.8),

$$F(z) - B_n(z) = z^{n+1} \prod_{k=0}^{n} (1- \mid \gamma_k \mid^2) \frac{F_{n+1}(z) - 1}{(\mathcal{B}_n(z) + z\mathcal{A}_n^T(z) F_{n+1}(z))(\mathcal{B}_n(z) + z\mathcal{A}_n^T(z))}.$$

■

1.2 Some classical extrapolation problems

In this section we relate the Schur algorithm to several classical extrapolation problems. The first one described here is the Schur problem:

'Given the complex numbers $\{c_k\}_{k=0}^n$, it is required to find conditions for the existence of a function F in $\mathcal{S}$ such that the given numbers are the first $n+1$ Taylor coefficients of F.'

A necessary condition can be easilsy obtained by considering the Toeplitz operator of the symbol F. Let H^2 be the Hardy space of analytic functions on $\mathbf{D}$ satisfying

$$\| G \|_{H^2} = \sup_{0<r<1} \frac{1}{2\pi} \int_0^{2\pi} \mid G(re^{it}) \mid^2 dt < \infty.$$

Then H^2 is a Hilbert space and for $F \in H^\infty$ we define

$$T_F G = P_{H^2}^{L^2} F G$$

where $P_{H^2}^{L^2}$ is the orthogonal projection of L^2 onto H^2.

T_F is bounded and $\| T_F \| = \| F \|_\infty = \sup_{z \in \mathbf{D}} \mid F(z) \mid$. As the family $\chi_n(z) = z^n$, $n \geq 0$, is an orthogonal basis of H^2, T_F has the matrix

$$\begin{bmatrix} c_o & 0 & 0 \cdots \\ c_1 & c_o & 0 \cdots \\ c_2 & c_1 & c_o \cdots \\ \cdot & \cdot & \cdot \\ \cdot & \cdot & \cdot \\ \cdot & \cdot & \cdot \\ c_n & c_{n-1} & \cdots \end{bmatrix}$$

with respect to this basis, $\{c_n\}_{n\geq 0}$ being the Taylor coefficients of F. We can now give a solution of the Schur problem.

Theorem 1.4 The Schur problem is solvable if and only if the matrix

$$T_n = \begin{bmatrix} c_o & 0 & 0\cdots 0 \\ c_1 & c_o & 0\cdots 0 \\ \cdot & \cdot & \cdot \\ \cdot & \cdot & \cdot \\ \cdot & \cdot & \cdot \\ c_n & c_{n-1} & c_{n-2}\cdots c_o \end{bmatrix}$$

built up on the given numbers $\{c_k\}_{k=0}^n$ is a contraction (i.e. $\| T_n \| \leq 1$).

Proof If F is a solution of the problem, $\| T_F \| = \| F \|_\infty \leq 1$ and T_n is the compression of T_F to the subspace generated by $\{\chi_k\}_{k=0}^n$; therefore T_n is a contraction.

We prove the sufficiency part by induction. The first step is obvious: if $T_o = [c_o]$ is a contraction, $| c_o | \leq 1$ and we define $F \equiv c_o$. We suppose now that the theorem is valid for all k, $0 \leq k \leq n-1$. For an analytic function on $\mathbf{D}$, $F(z) = c_o + c_1 z + c_2 z^2 + \cdots$, we define;

$$T_n(F) = \begin{bmatrix} c_o & 0 & 0\cdots 0 \\ c_1 & c_o & 0\cdots 0 \\ \cdot & \cdot & \cdot \\ \cdot & \cdot & \cdot \\ \cdot & \cdot & \cdot \\ c_n & c_{n-1} & c_{n-2}\cdots c_o \end{bmatrix} \tag{1.11}$$

and we have the properties;

$$T_n(\lambda_1 F_1 + \lambda_2 F_2) = \lambda_1 T_n(F_1) + \lambda_2 T_n(F_2) \tag{1.12}$$

$$T_n(F_1 F_2) = T_n(F_1) T_n(F_2). \tag{1.13}$$

Now take

$$T_n = \begin{bmatrix} c_o & 0 & 0\cdots 0 \\ c_1 & c_o & 0\cdots 0 \\ \cdot & \cdot & \cdot \\ \cdot & \cdot & \cdot \\ \cdot & \cdot & \cdot \\ c_n & c_{n-1} & c_{n-2}\cdots c_o \end{bmatrix}$$

to be a contraction and define

$$T_n' = (T_n - c_o I_{n+1})(I_{n+1} - \bar{c}_o T_n)^{-1}$$

(of course, we suppose that $| c_o |< 1$). We have that

$$I - T_n'^* T_n' = (1 - | c_o |^2)(I - c_o T_n^*)^{-1}(I - T_n^* T_n)(I - \bar{c}_o T_n)^{-1}$$

and so T_n' is a contraction. But, by its definition,

$$T_n' = \begin{bmatrix} 0 & 0 & \cdots 0 \\ c_o' & 0 & \cdots 0 \\ c_1' & c_o' & \cdots 0 \\ \cdot & \cdot & \cdot \\ \cdot & \cdot & \cdot \\ \cdot & \cdot & \cdot \\ c_{n-1}' & c_{n-2}' & \cdots c_o' \end{bmatrix}.$$

Consequently

$$\begin{bmatrix} c_o' & 0 & \cdots 0 \\ c_1' & c_o' & \cdots 0 \\ \cdot & \cdot & \cdot \\ \cdot & \cdot & \cdot \\ \cdot & \cdot & \cdot \\ c_{n-1}' & c_{n-2}' & \cdots c_o' \end{bmatrix}$$

is also a contraction and so, by our assumption, there exists a function $F \in \mathcal{S}$ having as the first n Taylor coefficients the numbers $c_o', \cdots, c_{n-1}'$. Take now

$$F(z) = \frac{c_o + zF'(z)}{1 + z\bar{c}_o F'(z)} \in \mathcal{S}.$$

By direct computations,

$$\begin{aligned} T_n(F) &= (c_o I_{n+1} + T_n(zF'))(I_{n+1} + z\bar{c}_o T_n(zF'))^{-1} \\ &= (c_o I_{n+1} + T_n')(I_{n+1} + \bar{c}_o T_n')^{-1} \\ &= T_n. \end{aligned}$$

This means that F is a solution of the problem with the given numbers $c_o, c_1, \cdots, c_n$. ∎

Equally important in classical extrapolation theory is the study of the set of the solutions of a given problem. Take $c_o, c_1, \cdots, c_n$ such that the associated T_n is a contraction, i.e. the Schur problem has a solution. In view of Theorem 1.1, for any solution, the first $n+1$ Schur parameters $\{\gamma_m\}_{m=0}^n$ are fixed and uniquely determined by $c_o, c_1, \cdots, c_n$. Consequently, for the polynomials $\mathcal{A}_n(\{\gamma_m\}_{m=0}^n)$ and $\mathcal{B}_n(\{\gamma_m\}_{m=0}^n)$ given by (1.3) and (1.4) we can also write $\mathcal{A}_n(\{c_m\}_{m=0}^n)$ and $\mathcal{B}_n(\{c_m\}_{m=0}^n)$.

Theorem 1.5 Consider the solvable Schur problem with the given numbers $c_o, c_1, \cdots, c_n$. Either this Schur problem has a unique solution, or its solutions are given by:

$$F(z) = \frac{\mathcal{A}_n(\{c_m\}_{m=0}^n) + z\mathcal{B}_n^T(\{c_m\}_{m=0}^n)E(z)}{\mathcal{B}_n(\{c_m\}_{m=0}^n) + z\mathcal{A}_n^T(\{c_m\}_{m=0}^n)E(z)} \tag{1.14}$$

with $E \in \mathcal{S}$.

Proof If there exists $n_o \leq n$ such that $|\ \gamma_{n_0}\ | = 1$, then the corresponding Schur problem has only one solution. Otherwise, any completion of $\gamma_o, \cdots, \gamma_n$ to a sequence of Schur parameters gives us a solution of the problem. The formula (1.2) and Remark 1.2 now give the representation (1.14) the set for all solutions.

There are now a few problems which can be handled using the solution of the Schur problem. First, we take into consideration the Carathéodory problem:

'Given the complex numbers $\{\frac{s_o}{2}, s_1, \cdots, s_n\}$, $s_o > 0$, it is required to determine conditions for the existence of a function G in the Carathéodory class of analytic functions in $\mathbf{D}$ with positive real part (and we suppose also that $G(0) > 0$), such that the given numbers are the first $n+1$ Taylor coefficients of G'.

We denote by $\mathcal{C}$ the Carathéodory class and there is a close connection between $\mathcal{C}$ and $\mathcal{S}$: $G \in \mathcal{C}$ if and only if

$$F(z) = \frac{G(z) - G(0)}{z(G(z) + G(0))}$$

belongs to $\mathcal{S}$.

Theorem 1.6 The Carathéodory problem is solvable if and only if the matrix

$$M_n = \begin{bmatrix} s_o & s_1 & \cdots s_n \\ \bar{s}_1 & s_o & \cdots s_{n-1} \\ \cdot & \cdot & \cdot \\ \cdot & \cdot & \cdot \\ \cdot & \cdot & \cdot \\ \bar{s}_n & \bar{s}_{n-1} & \cdots s_o \end{bmatrix} \tag{1.15}$$

built up on the given numbers $\{\frac{s_o}{2}, s_1, \cdots, s_n\}$ is positive semi-definite.

Proof If $G(z) = \frac{s_o}{2} + s_1 z + \cdots + s_n z^n + \cdots \in \mathcal{C}$ and $F(z) = c_o + c_1 z + \cdots + c_n z^n + \cdots \in \mathcal{S}$ are connected by the formula

$$F(z) = \frac{G(z) - G(0)}{z(G(z) + G(0))}$$

then their coefficients satisfy the following relations:

$$\begin{aligned} s_1 &= c_o s_o \\ &\cdot \\ &\cdot \\ &\cdot \\ s_n &= s_o c_{n-1} + s_1 c_{n-2} + \cdots + s_{n-1} c_o. \end{aligned} \tag{1.16}$$

This shows that to fix the first $n+1$ Taylor coefficients of G means to fix the first $n+1$ Taylor coefficients of F. Moreover, with the notation in the proof of Theorem 1.4 and using (1.12) and (1.13), we get:

$$T_n(G/G(0)) = (1 - T_n(zF))(I + T_n(zF))^{-1}.$$

$$T_n(G/G(0)) + T_n(G/G(0))^* = 2(I + T_n(zF)^*)^{-1}(I - T_n(zF)^* T_n(zF))(1 + T_n(zF))^{-1}$$

and

$$M_n = T_n(G) + T_n(G)^*.$$

These relations and Theorem 1.4 finish the proof. ∎

Theorem 1.7 If $M_n \geq 0$, then either the Carathéodory problem has a unique solution, or all solutions are given by:

$$G(z) = G(0) \frac{\hat{\psi}_n^T(z) + z\hat{\psi}_n(z)E(z)}{\hat{\phi}_n^T(z) - z\hat{\phi}_n(z)E(z)} \tag{1.17}$$

where $E \in \mathcal{S}$ and $\hat{\phi}_n, \hat{\psi}_n$ are monic polynomials of degree n, depending only on $\{s_1, \cdots, s_n\}$.

Proof The alternative in the statement of the theorem follows from Theorem 1.5. Also based on Theorem 1.5, we have the formula:

$$G(z) = G(0) \frac{\mathcal{B}_{n-1}(z) + z\mathcal{A}_{n-1}(z) + z(\mathcal{A}_{n-1}^T(z) + z\mathcal{B}_{n-1}^T(z))E(z)}{\mathcal{B}_{n-1}(z) - z\mathcal{A}_{n-1}(z) - z(z\mathcal{B}_{n-1}^T(z) - \mathcal{A}_{n-1}^T(z))E(z)}$$

for all the solutions of the Carathéodory problem. Denoting

$$\hat{\psi}_n(z) = z\mathcal{B}^T_{n-1}(z) + \mathcal{A}^T_{n-1}(z) \tag{1.18}$$

$$\hat{\phi}_n(z) = z\mathcal{B}^T_{n-1}(z) - \mathcal{A}^T_{n-1}(z) \tag{1.19}$$

we obtain (1.17). Moreover, having the properties of the polynomials $\mathcal{A}_n$ and $\mathcal{B}_n$, we find that $\hat{\phi}_n$ and $\hat{\psi}_n$ are monic and of degree n, depending only on $\{s_1, \cdots, s_n\}$. ∎

We pass on to the following trigonometric moment problem:

'Given the complex numbers $\{s_o, s_1, \cdots, s_n\}, s_o > 0$, it is required to find conditions for the existence of a positive finite Borel measure μ on the unit circle $\mathbf{T}$ such that the given numbers are the first $n+1$ Fourier coefficients of μ, i.e.

$$\frac{1}{2\pi}\int_0^{2\pi} e^{-ikt}d\mu(t) = s_k, \quad k = 0, 1, \cdots, n.'$$

Having in mind the Poisson representation of the harmonic functions on $\mathbf{D}$ and a compactness argument, we deduce the additive representation of the functions in the class $\mathcal{C}$:

$$G(z) = \frac{1}{4\pi}\int_0^{2\pi} \frac{e^{it}+z}{e^{it}-z}d\mu(t) \tag{1.20}$$

with μ a finite positive measure on $\mathbf{T}$. In this representation the correspondence between G and μ is one to one. Moreover, it results from (1.20) that

$$G(z) = \frac{1}{2\pi}\int_0^{2\pi} \left(\frac{1}{2} + \sum_{k=1}^{\infty} z^k e^{-ikt}\right)d\mu(t) \tag{1.21}$$

which shows that to fix the first $n+1$ Fourier coefficients of μ is the same as fixing $n+1$ Taylor coefficients of G.

Now, we can deduce the solution of the trigonometric moment problem.

Theorem 1.8 The trigonometric moment problem is solvable if and only if the matrix

$$M_n = \begin{bmatrix} s_o & s_1 & \cdots s_n \\ \bar{s}_1 & s_o & \cdots s_{n-1} \\ \cdot & \cdot & \cdot \\ \cdot & \cdot & \cdot \\ \cdot & \cdot & \cdot \\ \bar{s}_n & \bar{s}_{n-1} & \cdots s_o \end{bmatrix}$$

built up on the given numbers $\{s_o, s_1, \cdots s_n\}$ is positive semi–definite. Moreover, in this case, the problem either has a unique solution or a family of solutions parametrised by $\mathcal{S}$.

Proof We have to use only (1.20), Theorem 1.6 and Theorem 1.7. ∎

1.3 A Schur–type algorithm for operator–valued functions

The goal of this section is to point out a version of the Schur algorithm in the case of operator–valued functions.

We begin with some notations and definitions for operator–valued analaytic functions on the unit disc. Let us consider two Hilbert spaces $\mathcal{H}_1$ and $\mathcal{H}_2$, both separable, and let F be an analytic function on $\mathbf{D}$ with values in $\mathcal{L}(\mathcal{H}_1, \mathcal{H}_2)$, the set of bounded linear operators from $\mathcal{H}_1$ to $\mathcal{H}_2$. We denote by $\mathcal{S}(\mathcal{H}_1, \mathcal{H}_2)$ the class of such analytic functions satisfying $\| F(z) \| \leq 1$ for $z \in \mathbf{D}$.

It is known that F has a strong non–tangential limit a.e. on $\mathbf{T}$ as a consequence of Fatou's theorem and the separability of the spaces. In accordance with the duality (discussed in the Appendix), we define $\tilde{F} \in \mathcal{S}(\mathcal{H}_2, \mathcal{H}_1)$ by $\tilde{F}(z) = F(\bar{z})^*$, $z \in \mathbf{D}$ and we have that

$$\tilde{F}(e^{-it}) = F(e^{it})^* \quad \text{a.e.on } \mathbf{T}. \qquad 1.22$$

$F \in \mathcal{S}(\mathcal{H}_1, \mathcal{H}_2)$ is called inner if $F(e^{it})$ is an isometry a.e. on $\mathbf{T}$, *–inner if $\tilde{F}$ is inner, and inner from both sides if it is inner and *–inner in the same time.

We can also define the space $L^2(\mathcal{H})$ for a separable Hilbert space $\mathcal{H}$ as the classes of functions on $[0, 2\pi]$ with values in $\mathcal{H}$, measurable (strongly, or what is the same in view of the separability of $\mathcal{H}$, weakly), and satisfying the usual condition of L^2–boundedness:

$$\| g \|^2 = \frac{1}{2\pi} \int_o^{2\pi} \| g(t) \|_{\mathcal{H}}^2 \, dt < \infty.$$

If we identify functions in $L^2(\mathcal{H})$ which coincide almost everywhere with respect to Lebesgue measure, $L^2(\mathcal{H})$ becomes a Hilbert space. $H^2(\mathcal{H})$ is the set of analytic functions on $\mathbf{D}$ such that

$$\sup_{0 \leq r < 1} \frac{1}{2\pi} \int_0^{2\pi} \| G(re^{it}) \|_{\mathcal{H}}^2 \, dt < \infty$$

and it may be viewed as a subspece of $L^2(\mathcal{H})$.

For $F \in \mathcal{S}(\mathcal{H}_1, \mathcal{H}_2)$ we define the operator

$$T_f \in \mathcal{L}(H^2(\mathcal{H}_1), H^2(\mathcal{H}_2))$$
$$(T_F G)(z) = F(z)G(z), \quad G \in H^2(\mathcal{H}_1)$$

which is bounded. A function $F \in \mathcal{S}(\mathcal{H}_1, \mathcal{H}_2)$ is called outer if T_F has dense range; F is called *–outer if $\tilde{F}$ is outer.

Let us consider a contraction $T \in \mathcal{L}(\mathcal{H}_1, \mathcal{H}_2)$. i.e. $\| T \| \leq 1$. Define the defect operator and the defect space of T by $D_T = (I - T^*T)^{1/2}$ and $\mathcal{D}_T = \overline{\mathcal{R}(D_T)}$, where $\mathcal{R}$ will always mean the range of the underlying operator and $\bar{\mathcal{R}}$ is its closure. We have the relations

$$TD_T = D_{T^*}T \tag{1.23}$$

which follows from the equalities

$$T(D_T^2)^n = (D_{T^*}^2)^n T, \quad n = 0, 1, \cdots$$

and functional calculus for self–adjoint operators. The dual relation is

$$D_T T^* = T^* D_{T^*}. \tag{1.24}$$

From (1.23) and (1.24) we find that

$$T\mathcal{D}_T \subset \mathcal{D}_{T^*} \quad \text{and} \quad T^*\mathcal{D}_{T^*} \subset \mathcal{D}_T. \tag{1.25}$$

We have a canonical decomposition of contractions.

Lemma 1.9 For a contraction $T \in \mathcal{L}(\mathcal{H}_1, \mathcal{H}_2)$ we have the decomposition

$$T = \begin{bmatrix} T_p & 0 \\ 0 & T_u \end{bmatrix} : \begin{matrix} \mathcal{D}_T \\ \bigoplus \\ \ker D_T \end{matrix} \rightarrow \begin{matrix} \mathcal{D}_{T^*} \\ \bigoplus \\ \ker D_{T^*} \end{matrix}$$

where T_u is a unitary operator and T_p is a pure contraction (i.e. $\| T_p h \| < \| h \|$ for $h \neq 0$).

Proof To (1.25) we have to add the following remark:

$$\| D_T h \|^2 = \| h \|^2 - \| Th \|^2$$

which shows that $\ker D_T = \{h \in \mathcal{H}_1 \mid \| Th \| = \| h \|\}$. ∎

An application of the maximum principle yields the following variant of Lemma 1.9 for the class $\mathcal{S}(\mathcal{H}_1, \mathcal{H}_2)$:

Proposition 1.10 For $F \in \mathcal{S}(\mathcal{H}_1, \mathcal{H}_2)$ there exists the unlikely determined decomposition:

$$F(z) = \begin{bmatrix} F_p(z) & 0 \\ 0 & F_u \end{bmatrix} : \begin{matrix} \mathcal{D}_{F(0)} \\ \bigoplus \\ \ker D_{F(0)} \end{matrix} \rightarrow \begin{matrix} \mathcal{D}_{F(0)^*} \\ \bigoplus \\ \ker D_{F(0)^*} \end{matrix}$$

where $F_p \in \mathcal{S}(\mathcal{D}_{F(0)}, \mathcal{D}_{F(0)^*})$ has the property that $F_p(0)$ is a pure contraction (we will call F_p itself purely contractive) and F_u is a unitary constant. ∎

For the scalar Schur algorithm, an important role is played by the conformal maps of $\mathbf{D}$ onto $\mathbf{D}$, sometimes referred to as Möbius transformations. In an operatorial setting it is quite useful to replace Möbius transformations by the so–called cascade transformations.

Let $S = \begin{bmatrix} A & B \\ C & D \end{bmatrix} \in \mathcal{L}(\mathcal{H}_1 \bigoplus \mathcal{K}_1, \mathcal{H}_2 \bigoplus \mathcal{K}_2)$ be a contraction and define the cascade transformation of an arbitrary contraction $X \in \mathcal{L}(\mathcal{K}_2, \mathcal{K}_1)$ as

$$C_S(X) = A + BX(I - DX)^{-1}C \tag{1.26}$$

whenever the inverse of $(I - DX)$ exists. It only needs a little algebra to show that $C_S(X)$ is a contraction, and in a similar way we can define the cascade transformation for operator–valued functions.

The following functions will play a special role in the sequal. Fix $T \in \mathcal{L}(\mathcal{H}_1, \mathcal{H}_2)$ to be a contraction and define:

$$J_T(z) = \begin{bmatrix} T & zD_{T^*} \\ D_T & -zT^* \end{bmatrix} : \begin{matrix} \mathcal{H}_1 \\ \bigoplus \\ \mathcal{D}_{T^*} \end{matrix} \rightarrow \begin{matrix} \mathcal{H}_2 \\ \bigoplus \\ \mathcal{D}_T \end{matrix}. \tag{1.27}$$

It is quite easy to see that J_T is inner from both sides.

For $z = 1$, $J_T(1) = J(T)$ is the elementary rotation of T; see the Appendix. The main technical result in this section is the following.

Proposition 1.11 For any $F \in \mathcal{S}(\mathcal{H}_1, ' \iota_2)$ the equation

$$F = C_{J_{F(0)}}(G) \tag{1.28}$$

has a unique solution in $\mathcal{S}(\mathcal{D}_{F(0)}, \mathcal{D}_{F(0)^*})$.

Proof For the uniqueness of the solution of the given equation, we remark that if G is such a solution, then:

$$I - F(0)^* F(z) = D_{F(0)}(I + zF(0)^* G(z))^{-1} D_{F_{(0)}}. \tag{1.29}$$

From (1.29) we deduce that for two solutions G_1 and G_2 we have:

$$(I + zF(0)^* G_1(z))^{-1} = (I + zF(0)^* G_2(z))^{-1}. \tag{1.30}$$

On the other hand, as we have $C_{J_{F}(0)}(G_1) = C_{J_{F}(0)}(G_2)$, we obtain:

$$G_1(z)(1 + zF(0)^* G_1(z))^{-1} = G_2(z)(I + zF(0)^* G_2(z))^{-1}. \tag{1.31}$$

As a consequence of (1.30) and (1.31), $G_1 = G_2$.

For proving the existence of the solution of equation (1.28) we use an approximation procedure. From the very beginning we remark that as a consequence of the maximum principle, the operator $I - F(0)^* F(z)$ is one to one on $\mathcal{D}_{F(0)}$ for every fixed $z \in \mathbf{D}$. Also $I - F(z)^* F(0)$ is one to one on $\mathcal{D}_{F(0)}$ and consequently:

$$\overline{(I - F(0)^* F(z))\mathcal{D}_{F(0)}} = \mathcal{D}_{F(0)}. \tag{1.32}$$

Furthermore, as $F \in \mathcal{S}(\mathcal{H}_1, \mathcal{H}_2)$ it turns out that for any $n \geq 1$, the matrices $\begin{bmatrix} F(0) & 0 \\ C_n & F(0) \end{bmatrix}$ are contractions, where C_n are the Taylor coefficients of F. By Theorem A.7 in the Appendix, there exists a contraction $C'_n \in \mathcal{L}(\mathcal{D}_{F(0)}, \mathcal{D}_{F(0)^*})$ such that $C_n = D_{F(0)^*} C'_n D_{F(0)}$. We can compute:

$$\begin{aligned} I - F(0)^* F(z) &= I - F(0)^* F(0) - zF(0)^* D_{F(0)^*} C'_1 D_{F(0)} - \cdots - z^n F(0)^* D_{F(0)^*} C'_n D_{F(0)} - \cdots \\ &= D_{F(0)}(I - zF(0)^* C'_1 - \cdots - z^n F(0)^* C'_n - \cdots) D_{F(0)}. \end{aligned}$$

Since $D_{F(0)}$ is one to one on $\mathcal{D}_{F(0)}$ we see that the function

$$H_1(z) = I - zF(0)^* C'_1 - \cdots z^n F(0)^* C'_n - \cdots$$

is analytic in $\mathbf{D}$ with values in $\mathcal{L}(\mathcal{D}_{F(0)})$. We also define:

$$H(z) = H_1(z) D_{F(0)}$$

which is still analytic in $\mathbf{D}$. As $I - F(0)^*F(z) = D_{F(0)}H(z)$, we infer from (1.32) that $\overline{\mathcal{R}(H(z))} = \mathcal{D}_F(0)$. Moreover,

$$\overline{\mathcal{R}(H_1(z))} = \mathcal{D}_{F(0)}. \tag{1.33}$$

In view of (1.32) and the fact that $I - F(0)^*F(z)$ is one to one on $\mathcal{D}_{F(0)}$ we can define the generalised inverse of $I - F(0)^*F(z)$ on $\mathcal{D}_{F(0)}$ for any fixed $z \in \mathbf{D}$.

The main point of the proof is to define for a fixed $z \in \mathbf{D}$,

$$-F(0) + D_{F(0)^*}F(z)(I - F(0)^*F(z))^{-1}D_{F(0)} : \mathcal{R}(H(z)) \to \mathcal{D}^*_{F(0)}$$

by the formula:

$$\begin{aligned}(-F(0) &+ D_{F(0)^*}F(z)(I - F(0)^*F(z))^{-1}D_{F(0)})H(z)h \\ &= F(0)H(z)h + D_{F(0)^*}F(z)h, \quad h \in \mathcal{D}_{F(0)}.\end{aligned}$$

We claim that this operator is a contraction on $\mathcal{R}(H(z))$ and then it can be extended to the whole $\mathcal{D}_{F(0)}$ in view of the density of the range of $H(z)$. For $h \in \mathcal{D}_{F(0)}$,

$$\begin{aligned}&\| (-F(0) + D_{F(0)^*}F(z)(I - F(0)^*F(z))^{-1}D_{F(0)})H(z)h \|^2 \\ &=\| -F(0)H(z)h + D_{F(0)^*}F(z)h \|^2 \\ &=\| F(0)H(z)h \|^2 + \| D_{F(0)^*}F(z)h \|^2 - 2Re(F(0)H(z)h, D_{F(0)^*}F(z)h) \\ &=\| F(0)H(z)h \|^2 + \| F(z)h \|^2 - \| F(0)^*F(z)h \|^2 \\ &- 2Re(D_{F(0)}H(z)h, F(0)^*F(z)h) \\ &\leq\| F(0)H(z)h \|^2 - 2\,\mathrm{Re}(h, F(0)^*F(z)h) + \| F(0)^*F(z)h \|^2 + \| h \|^2 \\ &=\| F(0)H(z)h \|^2 + \| (I - F(0)^*F(z))h \|^2 \\ &=\| H(z)h \|^2 .\end{aligned}$$

In this way we obtain for every $z \in \mathbf{D}$ a contraction $G'(z)$ in $\mathcal{L}(\mathcal{D}_{F(0)}, \mathcal{D}_{F(0)^*})$. We will show that G' is analytic in $\mathbf{D}$ as a function of z. With this aim we compute for $h \in \mathcal{D}_{F(0)}$:

$$\begin{aligned}(I + F(0)^*G'(z))H(z)h &= H(z)h - F(0)^*F(0)H(z)h + F(0)^*D_{F(0)^*}F(z)h \\ &= D^2_{F(0)}H(z)h + D_{F(0)}F(0)^*F(0)h \\ &= D_{F(0)}((I - F(0)^*F(z))h + F(0)^*F(z)h) \\ &= D_{F(0)}h,\end{aligned}$$

which shows that

$$(I + F(0)^* G'(z)) H_1(z) = I.$$

But then, $H_1(z)$ will be invertible for $z \in \mathbf{D}$ and

$$H_1(z)^{-1} = I + F(0)^* G'(z). \tag{1.34}$$

From (1.34) $\| H_1(z)^{-1} \| \leq 2$ so we find that H_1^{-1} is an analytic function on $\mathbf{D}$. For $h \in \mathcal{H}_1$, we also have:

$$\begin{aligned}
&(F(0) + D_{F(0)^*} G'(z)(I + F(0)^* G'(z))^{-1} D_{F(0)})h \\
&= F(0)h + D_{F(0)^*} G'(z) H_1(z) D_{F(0)} h \\
&= F(0)h + D_{F(0)^*} G'(z) H(z) h \\
&= F(0)h - F(0) D_{F(0)} H(z) h + F(z)h - F(0)F(0)^* F(z) h \\
&= F(z)h.
\end{aligned}$$

From this relation we find that $G'(z)(I + F(0)^* G'(z))^{-1}$ is an analytic function on $\mathbf{D}$ and using (1.34),

$$G'(z) = G'(z)(I + F(0)^* G'(z))^{-1} H_1(z)$$

is an analytic function on $\mathbf{D}$. On the other hand, $G'(0) = 0$ and Schwartz's lemma gives us $G'(z) = zG(z)$ with $G \in \mathcal{S}(\mathcal{D}_{F(0)}, \mathcal{D}_{F(0)^*})$, so we have

$$F(z) = F(0) + z D_{F(0)^*} G(z)(I + zF(0)^* G(z))^{-1} D_{F(0)}$$

which means precisely that $C_{J_{F(0)}}(G) = F$.

Using Proposition 1.11 we can write an operatorial version of the Schur algorithm:

Operational Schur's algorithm Fix $F \in \mathcal{S}(\mathcal{H}_1, \mathcal{H}_2)$ and define:

$$F_o = F \quad \Gamma_n = F_n(0), \quad n \geq 0$$

$$F_{n+1} \in \mathcal{S}(\mathcal{D}_{\Gamma_n}, \mathcal{D}_{\Gamma_n^*}),$$

the unique solution of the equation

$$F_n = C_{J_{\Gamma_n}}(F_{n+1}).$$

By definition, $\{\Gamma_n\}_{n\geq 0}$ is a sequence of contractions, $\Gamma_o \in \mathcal{L}(\mathcal{H}_1, \mathcal{H}_2)$ and for $n \geq 1$, $\Gamma_n \in \mathcal{L}(\mathcal{D}_{\Gamma_{n-1}}, \mathcal{D}_{\Gamma^*_{n-1}})$. We continue to call these parameters Schur parameters of the function F. As a consequence of the algorithm, we obtain:

$$F = C_{J_{\Gamma_o}}(C_{J_{\Gamma_1}} \cdots (C_{J_{\Gamma_n}}(F_{n+1}) \cdots) \tag{1.35}P$$

and we have to use the so-called Redheffer product in order to write F as a single cascade transformation of F_{n+1}. Consequently, let us recall the assumptions involved in the Redheffer product. Consider two operator matrices $S = \begin{bmatrix} A & B \\ C & D \end{bmatrix}, S_1 = \begin{bmatrix} A_1 & B_1 \\ C_1 & D_1 \end{bmatrix}$ and try to find $S_1 * S$ such that

$$C_{S_1}(C_S(X)) = C_{S_1 * S}(X)$$

for any X acting between underlying spaces. With some algebra we find

$$S_1 * S = \begin{bmatrix} A_1 + B_1 A(1 - D_1 A)^{-1} C_1 & B_1 A(I - D_1 A)^{-1} D_1 B + B_1 B \\ C(1 - D_1 A)^{-1} C_1 & C(1 - D_1 A)^{-1} D_1 B + D \end{bmatrix} \tag{1.36}$$

whenever the involved inverses exist. We note that $*$ is associative and $J = \begin{bmatrix} 0 & I \\ I & 0 \end{bmatrix} = J(0)$ is the neutral element. Moreover, if S_1 and S are contractions, then $S_1 * S$ is also a contraction and a fortiori if S_1 and S are isometries (respectively co-isometries) then $S_1 * S$ is an isometry (co-isometry). We will return to these properties in the next section.

Returning now to the Schur algorithm, we define the functions

$$\begin{bmatrix} a_n & b_n \\ c_n & d_n \end{bmatrix} = J_{\Gamma_o} * J_{\Gamma_1} * \cdots * J_{\Gamma_n} : \begin{matrix} \mathcal{H}_1 \\ \oplus \\ \mathcal{D}_{\Gamma^*_n} \end{matrix} \rightarrow \begin{matrix} \mathcal{H}_2 \\ \oplus \\ \mathcal{D}_{\Gamma_n}. \end{matrix} \tag{1.37}$$

Consequently,

$$F = C_{\begin{bmatrix} a_n & b_n \\ c_n & d_n \end{bmatrix}}(F_{n+1}) \tag{1.38}$$

and a_n, b_n, c_n, d_n satisfy the following relations:

$$\begin{aligned} a_{n+1}(z) &= a_n(z) + b_n(z)\Gamma_{n+1}(I - d_n(z)\Gamma_{n+1})^{-1} c_n(z) \\ b_{n+1}(z) &= z b_n(z)(I - \Gamma_{n+1} d_n(z))^{-1} D_{\Gamma^*_{n+1}} \\ c_{n+1}(z) &= D_{\Gamma_{n+1}}(I - d_n(z)\Gamma_{n+1})^{-1} c_n(z) \\ d_{n+1}(z) &= z(-\Gamma^*_{n+1} + D_{\Gamma_{n+1}} d_n(z)(I - \Gamma_{n+1} d_n(z))^{-1} D_{\Gamma^*_{n+1}}). \end{aligned} \tag{1.39}PP$$

Moreover, as J_{Γ_n} are inner from both sides, we find that $\begin{bmatrix} a_n & b_n \\ c_n & d_n \end{bmatrix}$ are inner from both sides. We can prove now a result which extends Theore 1.1 to the operatorial Schur algorithm.

Theorem 1.12 The operatorial Schur algorithm realises a one to one correspondence between $\mathcal{S}(\mathcal{H}_1, \mathcal{H}_2)$ and the set of all $(\mathcal{H}_1, \mathcal{H}_2)$–Schur sequences (i.e. the sequences of contractions $\{\Gamma_n\}_{n\geq 0}$, $\Gamma_o \in \mathcal{L}(\mathcal{H}_1, \mathcal{H}_2)$, $\Gamma_n \in \mathcal{L}(\mathcal{D}_{\Gamma_{n-1}}, \mathcal{D}_{\Gamma^*_{n-1}})$.

Proof We have that

$$\begin{aligned} F(z) - a_n(z) &= C_{\begin{bmatrix} a_{n-1} & b_{n-1} \\ c_{n-1} & d_{n-1} \end{bmatrix}}(F_{\hat{n}})(z) - C_{\begin{bmatrix} a_{n-1} & b_{n-1} \\ c_{n-1} & d_{n-1} \end{bmatrix}}\Gamma_n(z) \\ &= b_{n-1}(z)(F_n(z)(I - d_{n-1}(z)F_n(z))^{-1} - \Gamma_n(I - d_{n-1}(z)\gamma_n)^{-1}c_{n-1}(z). \end{aligned}$$

But from (1.39)

$$b_n(z) = z^{n+1}v_n(z) \tag{1.40}$$

with $v_n \in \mathcal{S}(\mathcal{D}_{\Gamma^*_n}, \mathcal{H}_2)$ and $v_n(0) \neq 0$, and as $F_n(0) = \Gamma_n$ it follows that

$$F(z) - a_n(z) = z^{n+1}\hat{a}_n(z) \tag{1.41}$$

where $\hat{a}_n$ is a bounded analytic function on $\mathbf{D}$. As a consequence of (1.41) $\{a_n\}_{n\geq 0}$ converges uniformly on compact subsets of $\mathbf{D}$, in the uniform norm, to F. But, a_n are uniquely determined by $\{\Gamma_k\}_{k=0}^n$, therefore F is uniquely determined by $\{\Gamma_n\}_{n\geq 0}$. The converse is obvious. ∎

Remark 1.13 We have similar results to those in Remark 1.2. First, the Schur parameters of $F_n \in \mathcal{S}(\mathcal{D}_{\Gamma_{n-2}}, \mathcal{D}_{\Gamma^*_{n-1}})$ are $\{\Gamma_{n_1}\Gamma_{n+2}, \cdots\}$. Indeed, $(F_n)_1$ is the unique solution of the equation $F_n = C_{J_{F_{n(0)}}}(G)$. But $F_n(0) = \Gamma_n$ and in view of the Schur algorithm, $G = F_{n+1}$, so for $k \geq 1$, $(F_n)_k = F_{n+k}$ which shows that $\{\Gamma_{n_1}\Gamma_{n+1}, \cdots\}$ are the Schur parameters of F_n. Then the Schur parameters of $a_n \in \mathcal{S}(\mathcal{H}_1, \mathcal{H}_2)$ are $\{\Gamma_o, \Gamma_1, \cdots, \Gamma_n, 0, \cdots\}$.

From this, we see by (1.39) and Theorem 1.12 that the functions with the Schur parameters $\{\Gamma_o, \Gamma_1, \cdots, \Gamma_n, 0, \cdots\}$ are exactly a_n. ∎

1.4 Discrete transmission lines

In the previous sections we encountered several operations which can be better understood by using some diagrams.

An operator $S = \begin{bmatrix} A & B \\ C & D \end{bmatrix} \in \mathcal{L}(\mathcal{H}_1 \oplus \mathcal{K}_1, \mathcal{H}_2 \bigoplus \mathcal{K}_2)$ is viewed as (the transfer matrix of) a two–port and $X \in \mathcal{L}(\mathcal{K}_2, \mathcal{K}_1)$ is viewed as (the transfer matrix of) a one–port. By coupling these ports according to Fig 1.1, i.e. if $o_2 = i$ and $i_2 = o$, we obtain a one–port $C_S(X)$ given by

$$C_S(X) = A + BX(I - DX)^{-1}C$$

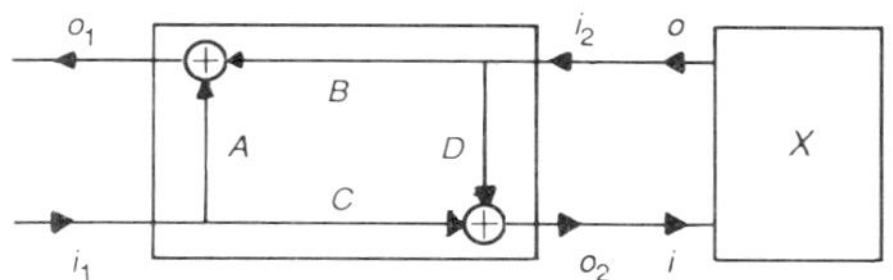

Fig 1.1

whenever the inverse of $I - DX$ exists. The formula is exactly the cascade transformation of X by S given by (1.26). In Fig 1.1 we may easily identify some properties of the cascade transformations; for instance, as we have already mentioned, if $\| D \| < 1$ and $\| S \| \leq 1$ then $\| C_S(X) \| \leq 1$. Indeed

$$\begin{aligned} \| i_1 \|^2 - \| C_S(X) i_1 \|^2 &= \| i_1 \|^2 - \| o_1 \|^2 \\ &= \| i_1 \|^2 + \| i_2 \|^2 - \| o_1 \|^2 - \| o_2 \|^2 + \| i \|^2 - \| o \|^2 \\ &= \| D_S \tbinom{i_1}{i_2} \|^2 + \| D_X i \|^2 . \end{aligned} \qquad (1.42)$$

We now pass on to the Redheffer product which replaces the composition of two cascade transformations by a single one.

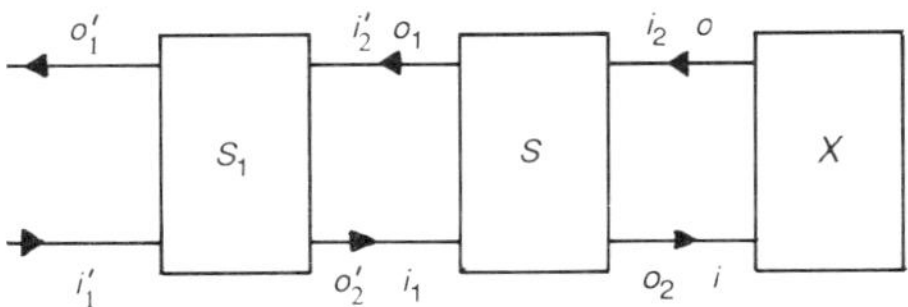

Fig 1.2

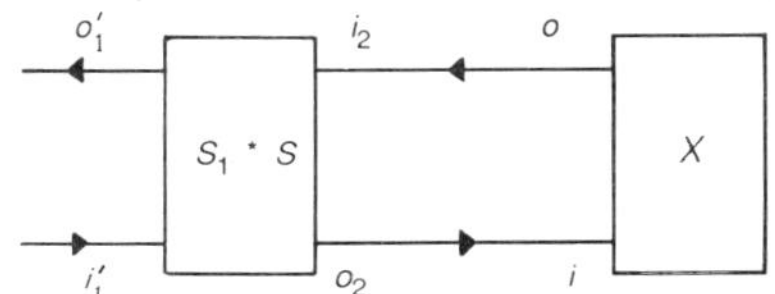

Fig 1.3

The illustration in Fig 1.2 is replaced by that in Fig 1.3 and we get exactly the formulas (1.36) for $S_1 * S$. Again some properties of the Redheffer product can be easily visualised. For instance, take $\| S_1 \| = 1$ and $\| S \| \leq 1$ with $\| A \| < 1$ then $\| S_1 * S \| \leq 1$ since

$$
\begin{aligned}
&\| \begin{pmatrix} i'_1 \\ i_2 \end{pmatrix} \|^2 - \| (S_1 * S) \begin{pmatrix} i'_1 \\ i_2 \end{pmatrix} \|^2 = \| i'_2 \|^2 + \| i_2 \|^2 - \| o^1_1 \|^2 - \| o \|^2 \\
&= (\| i'_1 \|^2 + \| i'_2 \|^2) - (\| o \|^2_1 + \| o'_1 \|^2) + (\| i_1 \|^2 + \| i_2 \|^2) - (\| o_2 \|^2 + \| o'_2 \|^2) \\
&= \| D_{S_1} \begin{pmatrix} i^1_1 \\ i^1_2 \end{pmatrix} \|^2 + \| D_S \begin{pmatrix} i_1 \\ i_2 \end{pmatrix} \|^2 .
\end{aligned}
$$

The visual interpretation of the Schur algorithm is now evident. If we are interested in the time evolution of a system, its transfer matrix will be an operatorial–valued analaytic bounded function on $\mathbf{D}$ (i.e. the system is linear, causal and with discrete time evolution). The first step in Schur's algorithm means the extraction of an elementary factor $J_{F(0)}$, such that we replace the one–port of Fig 1.4

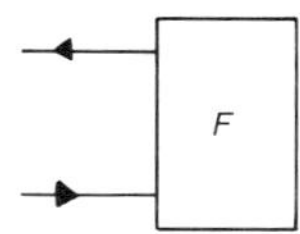

Fig 1.4

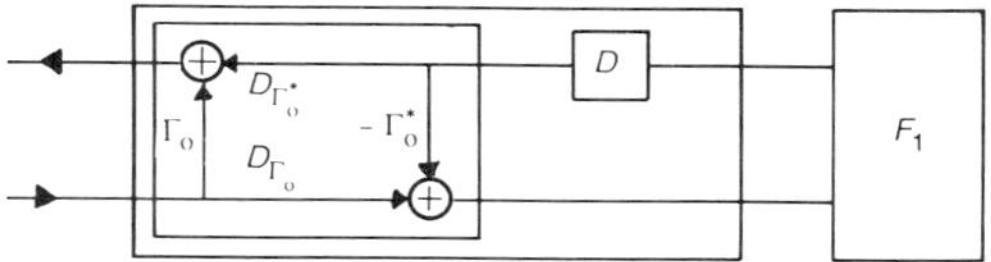

Fig 1.5

by that in Fig 1.5 where $\boxed{D}$ is the "time delay operator" caused by the presence of the factor z applied to $D_{\Gamma_o^*}$ and $-\Gamma_o^*$ in $J(\Gamma_o)$.

In this way the Schur algorithm is associated with a function $F \in \mathcal{S}(\mathcal{H}_1, \mathcal{H}_2)$ in Fig 1.6, usually called a discrete transmission line where $\{\Gamma_n\}_{n\geq 0}$ are the Schur parameters of the Function F.

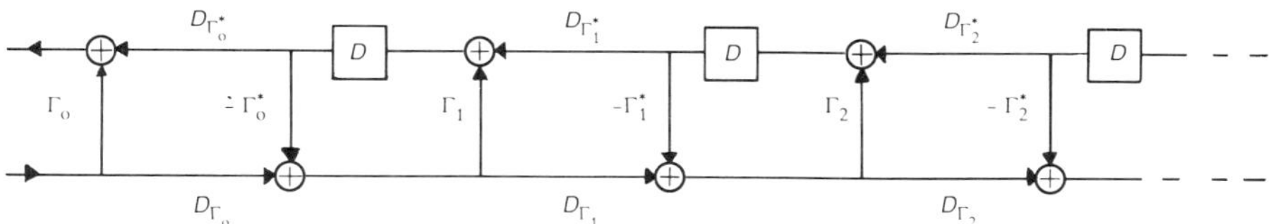

Fig 1.6

Notes

The results of Section 1.1 are from [78]. I. Schur's motivation came from the works of C. Carathéodory [18] and O. Toeplitz [85]. Section 1.2 illustrates how some classical extrapolation problems can be solved using Schur's method. For a general treatment, the monograph of M. G. Krein and A. A. Nudelman [61] is strongly recommended. A study of the operator–valued Schur class is undertaken in [70], [84], [75]. Proposition 1.11 and Theorem 1.12 are from the paper [27].

The history of extending Schur's ideas to an operatorial setting is rather long and we point out only that the first satisfactory notion extending that of Schur parameters appears in [20]. For Schur–type algorithms in the matricial case, we mention [31] and [35]. Section 1.4 includes some additional classical facts and here we followed [53] and [54] which may also be consulted for many other developments and references.

CHAPTER 2

Two completion algorithms

We have seen, especially in Section 1.2, that the idea involved in the Schur algorithm is closely connected with the completion of a certain structure. For instance, in the Schur problem we have to complete an $n \times n$ lower triangular Toeplitz contraction to an analaytic Toeplitz contraction. In the present chapter we give the details of such a completion procedure for lower triangular Toeplitz contractions and for positive Toeplitz matrices. Some other classical problems are presented as applications. Moreover, as a consequence of the analysis of positive (definite) Toeplitz matrices a construction of the Naimark dilation is indicated.

2.1 The structure of analytic Toeplitz contractions

As in the scalar case, any bounded analytic function on $\mathbf{D}$ with values in $\mathcal{L}(\mathcal{H}_1, \mathcal{H}_2)$ $\mathcal{H}_1, \mathcal{H}_2$ being Hilbert spaces, generates a lower triangular bounded Toeplitz operator acting between $\ell^2(\mathbf{N}) \bigotimes \mathcal{H}_1$ and $\ell^2(\mathbf{N}) \bigotimes \mathcal{H}_2$. This correspondence is also norm preserving. In this section we indicate a direct way of providing a certain structure for these lower triangular Toeplitz contractions. Here we will use most of the notation introduced in the Appendix. For instance, in the statement of the result below

$L_n = L_n(G_o, \cdots, G_n)$ is the row contraction given by Theorem A.4,

$\tilde{L}_n = \tilde{L}_n(G_o, \cdots, G_n)$ is the dual construction and, $V_n = V_n(G_o, \cdots, G_n)$ is the unitary operator connected with the elementary rotation of L_n by (A.2). Now, we can state and prove the main result of this section.

Theorem 2.1 There exists a one–to–one correspondence between the set of contractions

$$T_\infty = \begin{bmatrix} C_o & 0 & 0 \cdots \\ C_1 & C_o & 0 \cdots \\ C_2 & C_1 & C_o \cdots \\ \cdot & \cdot & \cdot \\ \cdot & \cdot & \cdot \\ \cdot & \cdot & \cdot \end{bmatrix}$$

on $\mathcal{L}(\ell^2(\mathbf{N})\bigotimes\mathcal{H}_1,\ell^2(\mathbf{N})\bigotimes\mathcal{H}_2)$ and the set of sequences of contractions $\{G_n\}_{n\geq 0}$, $G_o\in\mathcal{L}(\mathcal{H}_1,\mathcal{H}_2), G_k\in\mathcal{L}(\mathcal{D}_{G_{k-1}},\mathcal{D}_{G^*_{k-1}}), k\geq 1$, expressed by the formula

$$C_o = G_o$$

$$C_n = L_{n-1}Q_{n-2}\tilde{L}_{n-1} + D_{G^*_o}\cdots D_{G^*_{n-1}}G_nD_{G_{n-1}}\cdots D_{G_o}, \quad n\geq 1,$$

where $Q_{-1}=0:\mathcal{H}_1\to\mathcal{H}_2$ and for $n\geq 1$,

$$Q_n:\mathcal{H}_1\bigoplus\bigoplus_{k=0}^{n}\mathcal{D}_{G^*_k}\to\mathcal{H}_2\bigoplus\bigoplus_{k=0}^{n}\mathcal{D}_{G_k}$$

$$Q_n=(0\bigoplus I_n)V_n(Q_{n-1}\bigoplus I).$$

Proof We distinguish two main steps, the first one establishing the structure of the coefficients C_n and the second one analysing the operators Q_n.

Let us define the contractions

$$A_n=\begin{bmatrix} 0 & \cdots 0 & C_o \\ 0 & \cdots C_o & C_1 \\ \cdot & & \cdot \\ \cdot & & \cdot \\ \cdot & & \cdot \\ C_o & C_1\cdots & C_n \end{bmatrix}, \quad B_n=0\bigoplus A_n.$$

Moreover, we set $G_{-1}=0:\mathcal{H}_1\to\mathcal{H}_2$ and we prove by induction the existence of two sequences of unitary operators

$$\Omega_{n-1}:\mathcal{D}_{B_{n-1}}\to\bigoplus_{k=-1}^{n-1}\mathcal{D}_{G_k}$$

$$\tilde{\Omega}_{n-1}:\mathcal{D}_{B^*_{n-1}}\to\bigoplus_{k=-1}^{n-1}\mathcal{D}_{G^*_k}$$

together with the following properties:

$$(C_o,\cdots,C_n)^t = D_{B^*_{n-1}}\tilde{\Omega}^*_{n-1}\tilde{L}_n, \qquad (2.1)_n$$

$$(C_o,\cdots,C_n) = L_n\Omega_{n-1}D_{B_{n-1}}, \qquad (2.2)_n$$

there exists a uniquely determined contraction

$$G_{n+1} : \mathcal{D}_{G_n} \to \mathcal{D}_{G_n^*} \text{ with}$$

$$C_{n+1} = -L_n \Omega_{n-1} B_{n-1}^* \tilde{\Omega}_{n-1}^* \tilde{L}_n + D_{G_o^*} \cdots D_{G_n^*} D_{G_{n+1}} D_{G_n} \cdots D_{G_o}, \qquad (2.3)_n$$

$$\Omega_n : \mathcal{D}_{B_n} \to \bigoplus_{k=-1}^{n} \mathcal{D}_{G_k} \qquad (2.4)_n$$

satisfies

$$\Omega_n D_{B_n} = \begin{bmatrix} \Omega_{n-1} D_{B_{n-1}} & -\Omega_{n-1} B_{n-1}^* \tilde{\Omega}_{n-1}^* \tilde{L}_n \\ 0 & D_{G_n} \cdots D_{G_o} \end{bmatrix}$$

and

$$\tilde{\Omega}_n : \mathcal{D}_{B_n^*} \to \bigoplus_{k-1}^{n} \mathcal{D}_{G_k^*} \qquad (2.5)_n$$

satisfies

$$\tilde{\Omega}_n D_{B_n^*} = \begin{bmatrix} \tilde{\Omega}_{n-1} D_{B_{n-1}^*} & -\tilde{\Omega}_{n-1} B_{n-1} \Omega_{n-1}^* L_n^* \\ 0 & D_{G_n^*} \cdots D_{G_o^*} \end{bmatrix}.$$

For $n = 0$, $G_o = C_o$, $\Omega_o = I$ and $\tilde{\Omega}_o = I$. Let us give some details for the case $n = 1$. That is,

$$A_1 = \begin{bmatrix} 0 & C_o \\ C_o & C_1 \end{bmatrix}$$

and according to Theorem A.7, $C_1 = D_{G_o^*} G_1 D_{G_o}$, where $G_1 : \mathcal{D}_{G_o} \to \mathcal{D}_{G_o^*}$ is a uniquely determined contraction. Now, we have that

$$(C_o, C_1)^t = (G_o, D_{G_o^*} G_1 D_{G_o})^t = \begin{bmatrix} I & 0 \\ 0 & D_{G_o^*} \end{bmatrix} \tilde{L}_1 = D_{B_o^*} \tilde{\Omega}_o^* \tilde{L}_1 \qquad (2.1)_1$$

$$(C_o, C_1) = L_1 \begin{bmatrix} I & 0 \\ 0 & D_{G_o} \end{bmatrix} = L_1 \Omega_o D_{B_o} \qquad (2.2)_1$$

$$A_2 = \begin{bmatrix} B_o & D_{B_o}^* \tilde{\Omega}_o^* \tilde{L}_1 \\ L_1 \Omega_o D_{B_o} & C_2 \end{bmatrix} \qquad (2.3)_1$$

and again, by Theorem A.7,

$$\begin{aligned} C_2 &= -L_1 \Omega_o B_o^* \tilde{\Omega}_o^* \tilde{L}_1 + D_{L_1^*} G_2' D_{\tilde{L}_1} \\ &= -L_1 \Omega_o B_o^* \tilde{\Omega}_o^* \tilde{L}_1 + D_{G_o^*} D_{G_1^*} G_2 D_{G_1} D_{G_o} \end{aligned}$$

where we have also used Theorem A.4.

We can write

$$B_1 = \begin{bmatrix} 0 & 0 \\ B_o & D_{B_o^*} \tilde{\Omega}_o^* \tilde{L}_1 \end{bmatrix} \qquad (2.4)_1$$

and Ω_1 is defined according to the following commutative diagram

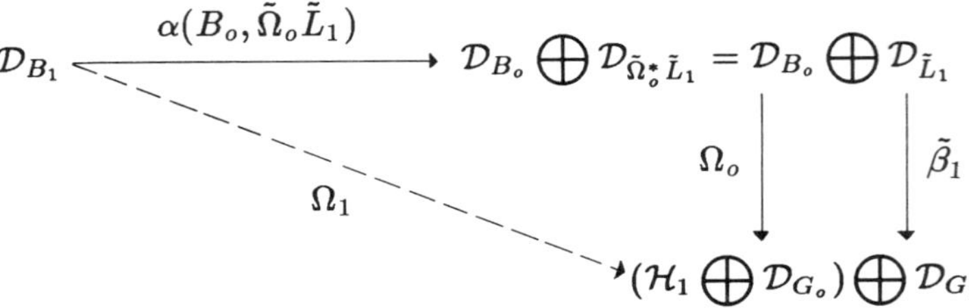

Define $\tilde{\Omega}_1$ in a similar way.

We can pass now to the general case: for $(2.1)_n$ we have

$$
\begin{aligned}
(C_o, \cdots, C_n)^t &= (D_{B^*_{n-2}} \tilde{\Omega}^*_{n-2} \tilde{L}_{n-1}, \\
&\quad - L_{n-1}\Omega_{n-2}B^*_{n-2}\tilde{\Omega}^*_{n-2}\tilde{L}_{n-1} + D_{G^*_o} \cdots G_n \cdots D_{G_o}) \\
&= \begin{bmatrix} D_{B^*_{n-2}}\tilde{\Omega}^*_{n-2} & 0 \\ -L_{n-1}\Omega_{n-2}B^*_{n-2}\tilde{\Omega}^*_{n-2} & D_{G^*_o} \cdots D_{G^*_{n-1}} \end{bmatrix} \begin{bmatrix} \tilde{L}_{n-1} \\ G_n D_{G_{n-1}} \cdots D_{G_o} \end{bmatrix} \\
&= D_{B^*_{n-1}} \tilde{\Omega}^*_{n-1} \tilde{L}_n.
\end{aligned}
$$

The proof of $(2.2)_n$ is similar to that of $(2.1)_n$. For $(2.3)_n$ we write

$$
A_{n+1} = \begin{bmatrix} B_{n-1} & D_{B^*_{n-1}} \tilde{\Omega}^*_{n-1} \tilde{L}_n \\ L_n \Omega_{n-1} D_{B_{n-1}} & C_{n+1} \end{bmatrix}
$$

and by Theorem A.7 and Theorem A.4,

$$
\begin{aligned}
C_{n+1} &= -L_n\Omega_{n-1}B^*_{n-1}\tilde{\Omega}^*_{n-1}\tilde{L}_n + D_{L^*_n} G'_{n+1} D_{\tilde{L}_n} \\
&= -L_n\Omega_{n-1}B^*_{n-1}\tilde{\Omega}^*_{n-1}\tilde{L}_n + D_{G^*_o} \cdots D_{G^*_n} G_{n+1} D_{G_n} \cdots D_{G_o}.
\end{aligned}
$$

For $(2.4)_n$, we can write

$$
B_n = \begin{bmatrix} 0 & 0 \\ B_{n-1} & D_{B^*_{n-1}} \tilde{\Omega}^*_{n-1} \tilde{L}_n \end{bmatrix}
$$

and we use the following commutative diagram for defining Ω_n:

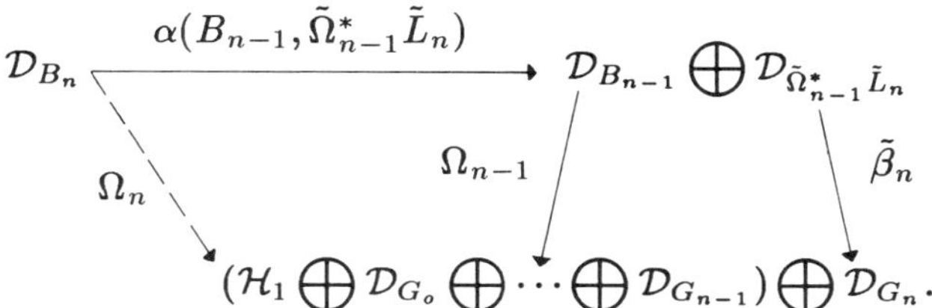

Finally, for $(2.5)_n$, we write

$$B_n = \begin{bmatrix} 0 & B_{n-1} \\ 0 & L_n \Omega_{n-1} D_{B_{n-1}} \end{bmatrix}$$

and we use the following commutative diagram

$$\begin{array}{ccc} \mathcal{D}_{B_n^*} & \xrightarrow{\tilde{\alpha}(B_{n-1}, L_n\Omega_{n-1})} & \mathcal{D}_{B_{n-1}^*} \bigoplus \mathcal{D}_{(L_n\Omega_{n-1})^*} \\ & \tilde{\Omega}_n \searrow & \tilde{\Omega}_{n-1} \downarrow \qquad\qquad \downarrow \beta_n \\ & & (\mathcal{H}_2 \bigoplus \mathcal{D}_{G_o^*} \bigoplus \cdots \bigoplus \mathcal{D}_{G_{n-1}^*} \bigoplus \mathcal{D}_{G_n^*} \end{array}$$

so that we can conclude the first step of the proof.

For the second step, we define $Q_{-1} = 0$ and for $n \geq 0$,

$$Q_n : \bigoplus_{k=-1}^{n} \mathcal{D}_{G_k^*} \to \bigoplus_{k=-1}^{n} \mathcal{D}_{G_k}$$

$$Q_n = -\Omega_n B_n^* \tilde{\Omega}_n^*. \tag{2.6}$$

First, we have

$$\begin{aligned} Q_o = -\Omega_o B_o^* \tilde{\Omega}_o^* &= -\begin{bmatrix} I & 0 \\ 0 & I \end{bmatrix} \begin{bmatrix} 0 & 0 \\ 0 & G_o^* \end{bmatrix} \begin{bmatrix} I & 0 \\ 0 & I \end{bmatrix} \\ &= \begin{bmatrix} 0 & 0 \\ 0 & I \end{bmatrix} \begin{bmatrix} G_o & D_{G_o^*} \\ D_{G_o} & -G_o^* \end{bmatrix} \begin{bmatrix} 0 & 0 \\ 0 & I \end{bmatrix} = (0 \bigoplus I_1) V_o (Q_{-1} \bigoplus I). \end{aligned}$$

For determining the structure of Q_n there is another way of identifying the defect space $\mathcal{D}_{B_n}$. Writing

$$B_n = \begin{bmatrix} 0 & B_{n-1} \\ 0 & L_n \Omega_{n-1} D_{B_{n-1}} \end{bmatrix}$$

the following commutative diagram can be taken into account:

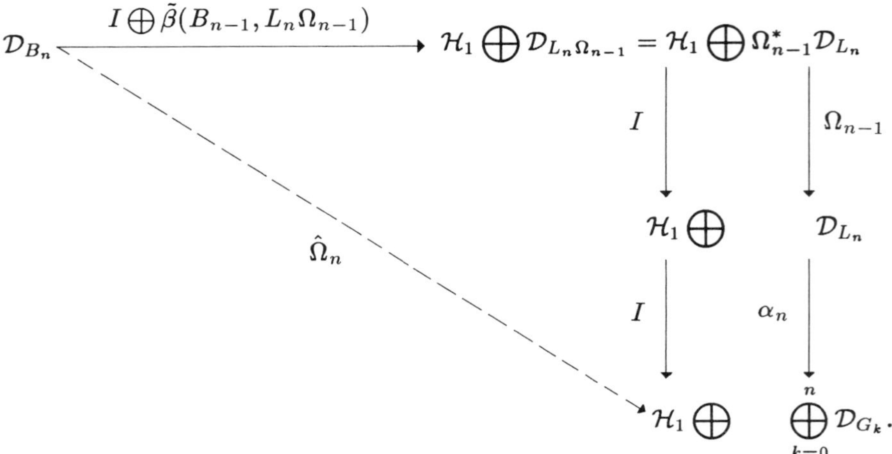

By defining $\Omega_n D_{B_n} = X_n$ and $\hat{\Omega}_n D_{B_n} = Y_n$, we have that $X_n^* X_n = Y_n^* Y_n$ and X_n, Y_n are upper triangular operators with dense ranges in $\bigoplus_{k=0}^{n} \mathcal{D}_{G_k}$.

By Proposition A.2, there exist unitary operators u_n such that $X_n = u_n Y_n$. But, X_n and Y_n have the same diagonal, and this forces $u_n = I$, i.e. $\Omega_n = \tilde{\Omega}_n$. Now, we can conclude the second step of the proof.

$$Q_n = -\Omega_n B_n^* \tilde{\Omega}_n^* = -\hat{\Omega}_n B_n^* \tilde{\alpha}^*(B_{n-1}, L_n\Omega_{n-1}) \begin{bmatrix} \tilde{\Omega}^*_{n-1} & 0 \\ 0 & \beta_n^* \end{bmatrix}$$
$$= -\hat{\Omega}_n (I \bigoplus \tilde{\beta}_n^*(B_{n-1}, L_n\Omega_{n-1}))$$
$$((I \bigoplus \tilde{\beta}_n(B_{n-1}.L_n\Omega_{n-1}))B_n^* \tilde{\alpha}_n^*(B_{n-1}, L_n\Omega_{n-1})) \begin{bmatrix} \tilde{\Omega}^*_{n-1} & 0 \\ 0 & \beta_n^* \end{bmatrix}$$

and taking Lemma A.8 into account,

$$Q_n = -\hat{\Omega}_n (I \bigoplus \tilde{\beta}_n^*(B_{n-1}, L_n\Omega_{n-1})) \begin{bmatrix} 0 & 0 \\ \Omega^*_{n-1} D_{L_n} & -\Omega^*_{n-1} L_n^* \beta_n^* \end{bmatrix} \begin{bmatrix} Q_{n-1} & 0 \\ 0 & I \end{bmatrix}$$
$$= \begin{bmatrix} I & 0 \\ 0 & \alpha_n \Omega_{n-1} \end{bmatrix} \begin{bmatrix} 0 & 0 \\ \Omega^*_{n-1} D_{L_n} & -\Omega^*_{n-1} L_n^* \beta_n^* \end{bmatrix} \begin{bmatrix} Q_{n-1} & 0 \\ 0 & I \end{bmatrix}.$$

Now, by the definition (A.2) of V_n, we get

$$Q_n = \begin{bmatrix} 0 & 0 \\ \alpha_n D_{L_n} & -\alpha_n L_n^* \beta_n^* \end{bmatrix} \begin{bmatrix} Q_{n-1} & 0 \\ 0 & I \end{bmatrix} = (0 \bigoplus I_n) V_n (Q_{n-1} \bigoplus I). \qquad \blacksquare$$

The parameters $G_n, n \geq 0$ appearing in Theorem 2.1, will be called the choice parameters of T_∞.

2.2 Connection with Schur's algorithm

Take a function $F \in \mathcal{S}(\mathcal{H}_1, \mathcal{H}_2)$ and ket $\{\Gamma_n\}_{n\geq 0}$ be its associated Schur parameters. Then, take $T_\infty = T_F$, the Toeplitz operator with symbol F and let $\{G_n\}_{n\geq 0}$ be the choice sequence of T_∞. In this way, we associate with F a uniquely determined choice sequence.

Proposition 2.2 If $F \in \mathcal{S}(\mathcal{H}_1, \mathcal{H}_2)$, $\{\Gamma_n\}_{n\geq 0}$ are its Schur parameters and $\{G_n\}_{n\geq 0}$ are its choice parameters, then

$$G_n = \Gamma_n, \quad n \geq 0.$$

Proof First of all we prove the following relations:

$$C_n = \text{formula } (\Gamma_o, \cdots, \Gamma_{n-1}) + D_{\Gamma_o^*} \cdots D_{\Gamma_{n-1}^*} \Gamma_n D_{\Gamma_{n-1}} \cdots D_{\Gamma_o}$$

where formula $(\Gamma_o, \cdots, \Gamma_n)$ is a formula depending only on the Schur parameters $\Gamma_o, \cdots,$ Γ_{n-1} and C_n are the Taylor coefficients of F. By induction and (1.39), we find that

$$c_n(0) = D_{\Gamma_n} \cdots D_{\Gamma_o} \tag{2.7}$$

and

$$(\frac{b_n}{z^{n+1}})(0) = D_{\Gamma_o^*} \cdots D_{\Gamma_n^*}. \tag{2.8}$$

Returning to the computation made in the proof of Theorem 1.12,

$$\begin{aligned} &F(z) - a_n(z) \\ &\quad = b_{n-1}(z)(F_n(z)(I - d_{n-1}(z)F_n(z))^{-1} - \Gamma_n(I - d_{n-1}(z)\Gamma_n)^{-1})c_{n-1}(z) \\ &\quad = b_{n-1}(z)(F_n(z) + F_n(z)d_{n-1}(z)F_n(z) + \cdots - \Gamma_n - \Gamma_n d_{n-1}(z)\Gamma_n - \cdots)c_{n-1}(z). \end{aligned}$$

From Remark 1.13, we have that $F_n(z) = \Gamma_n + zD_{\Gamma_n^*}\Gamma_{n+1}D_{\Gamma_n} + \cdots$. Taking into account that $d_n(0) = 0$, and that a_n depends only on $\Gamma_o, \cdots, \Gamma_n$ (2.7) and (2.8), we obtain

$$C_n = \text{ (the } n-\text{th Taylor coefficient of } a_{n-1}) + D_{\Gamma_o^*} \cdots D_{\Gamma_{n-1}^*} \Gamma_n D_{\Gamma_{n-1}} \cdots D_{\Gamma_o}.$$

We conclude that in both algorithms (Theorems 1.12 and 2.1) C_n lives in certain operatorial balls. As these balls coincide, they must have equal centres and equal left (right) radii. For more details, we proceed by induction. It is clear that $G_o = \Gamma_o$. Suppose $G_k = \Gamma_k$, $k = 0, \cdots, n-1$.

If we choose $\Gamma_n = 0$ there exists a contraction G_n^1 in $\mathcal{L}(\mathcal{D}_{G_{n-1}}, \mathcal{D}_{G_{n-1}^*})$ such that

$$\text{formula}\,(G_o, \cdots, G_{n-1}) + D_{G_o^*} \cdots G_n^1 \cdots D_{G_o} = L_{n-1} Q_{n-2} \tilde{L}_{n-1}.$$

For G_n^1 there exists a contraction G_n^2 in $\mathcal{L}(\mathcal{D}_{G_{n-1}}, \mathcal{D}_{G_{n-1}^*})$ such that

$$\begin{aligned}\text{formula}\,(G_o, \cdots G_{n-1}) &+ D_{G_o^*} \cdots G_n^2 \cdots D_{G_o} \\ &= L_{n-1} Q_{n_2} \tilde{L}_{n-1} + D_{G_o^*} \cdots G_n^1 \cdots D_{G_o}\end{aligned}$$

and we successively obtain k relations of this type. Adding up these equalities, we have

$$k L_{n-1} Q_{n-2} \tilde{L}_{n-1} = k \text{ formula}\,(G_o, \cdots, G_{n-1}) + D_{G_o^*} \cdots G_n^k \cdots D_{G_o}$$

or

$$L_{n-1} Q_{n-2} \tilde{L}_{n-1} - \text{ formula}\,(G_o, \cdots, G_{n-1}) = (\frac{1}{k})(D_{G_o^*} \cdots G_n^k \cdots D_{G_o})$$

and letting $k \to \infty$, we get

$$\text{formula}\,(G_o, \cdots, G_{n-1}) = L_{n-1} Q_{n-1} \tilde{L}_{n-1}$$

and we can immediately deduce that for any F, $G_n = \Gamma_n$.

Some supplementary information is contained in the proof of Theorem 2.1. First of all suppose that $\mathcal{H}_1$ and $\mathcal{H}_2$ have finite dimensions. As a consequence of the formulas $(4.4)_n$ and $(2.5)_n$, we get

$$\begin{aligned}\det(I - T_n^* T_n) &= \prod_{k=o}^{n} (\det D_{G_k})^{2(n-k+1)} \\ &= \prod_{k=o}^{n} (\det D_{G_k^*})^{2(n-k+1)} \\ &= \det(I - T_n T_n^*). \end{aligned} \tag{2.9}$$

Based on this formula, we can specify the case when the Schur problem has a unique solution.

Proposition 2.3 Let the complex numbers $\{c_k\}_{k=0}^n$ be given such that $T_n(\{c_k\}_{k=0}^n) \geq 0$. Then, the following assertions are eqivalent:

1. The Schur problem with data $\{c_k\}_{k=0}^n$ has a unique solution.
2. There exists $0 \leq r \leq n$ such that $\mid g_r \mid = 1$ ($\{g_k\}_{k=0}^n$ being the choice parameters of T_n, given by Theorem 2.1).
3. $\text{rank}(I - T_n^* T_n) = r$.
4. $\det(I - T_r^* T_r) = 0$ for some $0 \leq r \leq n$, but $\det(I - T_p^* T_p) \neq 0$ for $0 \leq p < r$.
5. $\det(I - T_n^* T_n) = 0$.

Now, we can solve another classical problem, known as the Carathéodory–Fejér problem

'Given the complex numbers$\{c_k\}_{k=0}^m (\sum_{k=0}^n \mid c_k \mid > 0)$ it is required to compute $m(\{c_k\}_{k=o}^n) = \inf \{\| F \|_\infty \mid F \in H^\infty$ and F has the first $n+1$ Taylor coefficients the given numbers $C_o, C_1, \cdots, C_n\}$.'

Theorem 2.4

$$m(\{c_k\}_{k=0}^n) = \lambda_M$$

where λ_M is the maximum root of the equation $\det(\lambda^2 I - T_n T_n) = 0$. The infimum in the Carathéodory–Fejér problem is actually a minimum attained for a unique function F^o of the form

$$F^o(z) = \epsilon \lambda_M \frac{P(z)}{P^T(z)}$$

where $\mid \epsilon \mid = 1$ and P is a polynomial of degree r=rank$(\lambda_M^2 I - T_n^* T_n)$.

Proof One inequality is obvious

$$\lambda_M^2 = \| T_n^* T_n \| = \| T_n \|^2 \leq \| T_F \|^2 = \| F \|_\infty^2$$

for any $F \in H^\infty$ having as the first $n+1$ Taylor coefficients the numbers $\{c_k\}_{k=0}^n$. For the converse, we can assume without loss of generality that $c_o \neq 0$, i.e. $\| T_n(\{c_k\}_{k=0}^n) \| \neq 0$. The matrix $T_n' = T_n(\{c_k / \| T_n(\{\frac{c_k}{\| T_n(\{c_k\}_{k=0}^n) \|}\}_{k=0}^n)$ has norm 1: consequently, $\det(I - T_n'^* T_n') = 0$ and in view of Proposition 2.3, there exists a unique function F' in

$\mathcal{S}$ having as its first $n+1$ Taylor coefficients, the numbers $\{\frac{c_k}{\| T_n(\{c_k\}_{k=0}^n \|}\}_{k=0}^n$ and its norm equal to 1. In view of Theorem 1.5

$$F'(z) = \frac{\mathcal{A}_{r-1}(z) + z\mathcal{B}_{r-1}^T(z)\epsilon}{\mathcal{B}_{r-1}(z) + z\mathcal{A}_{r-1}^T(z)\epsilon}$$

with a uniquely determined number ϵ of modulus 1 and $r = \text{rank}(I - T_n'^* T_n') = \text{rank}(\lambda_M^2 I - T_n^* T_n)$. Taking $F^o = \lambda_M F'$, we find that F^o has the numbers $\{c_k\}_{k=o}^n$ as its first $n+1$ Taylor coefficients and $\| F^o \|_\infty = \lambda_M$. Define $P(z) = \mathcal{A}_{r-1}(z) + z\mathcal{B}_{r-1}^T(z)\epsilon$; then $P^{\mathrm{T}}(z) = z\mathcal{A}_{r-1}^{\mathrm{T}}(z) + \bar{\epsilon}\mathcal{B}_{r-1}(z)$ and we have written F^o in the described form.

From Theorem 2.1 we deduce another simple solution to Schur's problem in the following operatorial formulation.

'Given the operators $C_k \in \mathcal{L}(\mathcal{H}_1, \mathcal{H}_2)$, $k = 0, \cdots, n$, it is required to find conditions for the existence of a function $F \in \mathcal{S}(\mathcal{H}_1, \mathcal{H}_2)$ such that the given operators are the first $n+1$ Taylor coefficients of F.'

Theorem 2.5 The operatorial Schur problem is solvable if and only if the block matrix

$$T_n = \begin{bmatrix} C_o & 0 & \cdots & 0 \\ C_1 & C_o & \cdots & 0 \\ . & . & . & . \\ C_n & C_{n-1} & \cdots & C_o \end{bmatrix}$$

built on the given operators $\{C_k\}_{k=0}^n$ is a contraction.

Proof One implication is immediate. For the converse, we suppose that T_n is a contraction. Then, the algorithm in Theorem 2.1 will produce a sequence of contractions $\{G_k\}_{k=0}^n, G_o = C_o \in \mathcal{L}(\mathcal{H}_1, \mathcal{H}_2)$ and $G_k \in \mathcal{L}(\mathcal{D}_{G_{k-1}}, \mathcal{D}_{G_{k-1}^*}), 1 \le k \le n$. As this sequence can always be extended to a choice sequence of infinite length again by Theorem 2.1 this choice sequence will produce a solution for the given Schur problem. ∎

Regarding the parametrisation of the solutions, we have a result which is similar to Theorem 1.5. Indeed, let T_n be a contraction (i.e. the Schur problem based on the data $C_o, \cdots, C_n$ is solvable) and let $\{G_k\}_{k=0}^n$ be the choice parameters associated with T_n by Theorem 2.1. With these parameters we can construct the functions a_n, b_n, c_n, d_n defined by (1.39) and in view of (1.38), Theorem 2.1 and Proposition 2.2, we obtain the following result.

Theorem 2.6 Consider a solvable operatorial Schur problem where the given data $C_o, \cdots, C_n$ are operators in $\mathcal{L}(\mathcal{H}_1, \mathcal{H}_2)$. Then, either this Schur problem has a unique solution (this situation arising precisely when one of its choice parameters G_k, $\leq k \leq n$ is an isometry of a co-isometry) or its solutions are given by the formula

$$F = C_{\begin{bmatrix} a_n & b_n \\ c_n & d_n \end{bmatrix}}(E) \tag{2.10}$$

with

$$E \in \mathcal{S}(\mathcal{D}_{G_n}, \mathcal{D}_{G_n^*}). \qquad \blacksquare$$

We can formulate another generalisation of the Schur problem as follows.

'Given operators $\{A_k\}_{k=0}^n \in \mathcal{L}(\mathcal{H}_3, \mathcal{H}_2)$ and $\{B_k\}_{k=0}^n \in \mathcal{L}(\mathcal{H}_1, \mathcal{H}_2)$, it is required to find conditions in order for $F \in \mathcal{S}(\mathcal{H}_1, \mathcal{H}_3)$ to exist such that

$$\sum_{i=0}^{j} A_{j-1} C_i = B_j, \quad j = 0, \cdots, n,$$

$\{C_n\}_{n=0}^\infty$ being the Taylor coefficients of F.'

In order to avoid some difficulties (and in order to have a simple parametrisation of the solutions), we suppose that $\ker A_o = 0$. We remark that in the scalar case ($\mathcal{H}_1, \mathcal{H}_2, \mathcal{H}_3$) are one-dimensional) this is not a restriction.

Theorem 2.7 Given $\{A_k\}_{k=0}^n \in \mathcal{L}(\mathcal{H}_3, \mathcal{H}_2)$ and $\{B_k\}_{k=0}^n \in \mathcal{L}(\mathcal{H}_1, \mathcal{H}_2)$ with $\ker A_o = 0$, the above-stated problem is solvable if and only if

$$T_n(\{A_k\}_{k=0}^n) T_n(\{A_k\}_{k=0}^n)^* \geq T_n(\{B_k\}_{k=0}^n) T_n(\{B_k\}_{k=0}^n)^*. \tag{2.11}$$

Moreover, if the problem is solvable, the operators $\{C_k\}_{k=0}^n$ are uniquely determined by $\{A_k\}_{k=0}^n$ and $\{B_k\}_{k=0}^n$, and the solutions of the problem are given by the solutions of the Schur problem with data $\{C_k\}_{k=0}^n$.

Proof The relation between $\{A_k\}_{k=0}^n$, $\{B_k\}_{k=0}^n$ and $\{C_k\}_{k=0}^n$ is equivalent to

$$T_n(\{A_k\}_{k=0}^n) T_n(\{C_k\}_{k=0}^n) = T_n(\{B_k\}_{k=0}^n$$

so that the necessity of the condition (2.11) is obvious. Now, suppose (1.11) is fulfilled. There exists a contraction $C = (C_{ij})_{i,j=0}^{n}$ such that

$$T_n(\{A_k\}_{k=0}^{n}C = T_n(\{B_k\}_{k=0}^{n}).$$

But, in view of our assumption that $\ker A_0 = 0$, it turns out that $\ker T_n(\{A_k\}_{k=0}^{n}) = 0$ and this shows that C is unique. Moreover, again by the assumption that $\ker A_0 = 0$, we find that C is a lower triangular contraction. From now on, we have only to use Theorem 2.6. ∎

In the scalar and matricial case, there exists a well–stated factorisation theory for functions in Schur classes. For the operatorial case there appear again several difficulties, and we end this section with some considerations of inner functions and finite 'Blaschke products' in the class $\mathcal{S}(\mathcal{H}_1, \mathcal{H}_2), \mathcal{H}_1, \mathcal{H}_2$ being separable Hilbert spaces.

Proposition 2.8 $F \in \mathcal{S}(\mathcal{H}_1, \mathcal{H}_2)$ is inner if and only if T_F is an isometry.

Proof If F is inner, then it is clear that T_F is an isometric operator. Conversely, if T_F is assumed to be an isometry, then

$$\frac{1}{2\pi}\int_0^{2\pi} \| g(t) \|_{\mathcal{H}_1}^2 \, dt = \frac{1}{2\pi}\int_0^{2\pi} \| F(e^{it})g(t) \|_{\mathcal{H}_2}^2 \, dt$$

for every $g \in L^2(\mathcal{H}_1)$. Choosing $g(t) = \chi_{(\tau,\tau+\delta)}(t)h$, with $h \in \mathcal{H}_1$ and $\chi_{(\tau,\tau+\delta)}$ the characteristic function of the interval $(\tau, \tau+\delta)$, we get

$$\frac{1}{\delta}\int_{\tau}^{\tau+\delta} \| F(e^{it})h \|_{\mathcal{H}_2}^2 \, dt = \| h \|_{\mathcal{H}_1}^2$$

which implies that

$$\| F(e^{it})h \|_{\mathcal{H}_2}^2 = \| h \|_{\mathcal{H}_1}^2$$

apart from a set of zero measure depending on h. Letting h run over a countable set dense in $\mathcal{H}_1$, we conclude that $F(e^{it})$ is an isometry for almost every t. ∎

Now fix $F \in \mathcal{S}(\mathcal{H}_1, \mathcal{H}_2)$ and let $\{G_n\}_{n=0}^{\infty}$ be its choice sequence. Define for $m \geq n$,

$$D_n^m : \underbrace{\mathcal{H}_1 \bigoplus \cdots \bigoplus \mathcal{H}_1}_{m-n} \bigoplus \mathcal{H}_1 \bigoplus \mathcal{D}_{G_0} \bigoplus \cdots \bigoplus \mathcal{D}_{G_{m-1}}$$

$$\to \underbrace{\mathcal{H}_1 \bigoplus \cdots \bigoplus \mathcal{H}_1}_{m-n} \bigoplus \mathcal{D}_{G_0} \bigoplus \cdots \bigoplus \mathcal{D}_{G_m}$$

$$D_n^m = \begin{bmatrix} I_{m-n} & 0 \\ 0 & D_n \end{bmatrix}. \qquad (2.12)$$

Lemma 2.9 For $n \geq 1$,

$$(C_o, C_1, \cdots, C_n) = L_n D_{n-I}^n \cdots D_o^n.$$

Proof By formula $(2.2)_n$ in the proof of Theorem 2.1,

$$(C_o, C_1, \cdots, C_n) = L_n \Omega_{n-1} D_{B_{n-1}}$$

so that, we have to prove the equality

$$\Omega_{n-1} D_{B_{n-1}} = D_{n-1}^n \cdots D_o^n.$$

But taking into account the second expression of Ω_{n-1} in the proof of Theorem 2.1,

$$\Omega_{n-1} D_{B_{n-1}} = \begin{bmatrix} I & 0 \\ 0 & D_{n-1} \end{bmatrix} \begin{bmatrix} I & 0 \\ 0 & \Omega_{n-2} D_{B_{n-2}} \end{bmatrix}$$

which is the main step of a simple induction. ∎

Define the unitary operators

$$u_n = \begin{bmatrix} 0 & \cdots 0 & I \\ 0 & \cdots I & 0 \\ \cdot & & \cdot \\ \cdot & & \cdot \\ \cdot & & \cdot \\ I & 0 & \cdots 0 \end{bmatrix} \qquad (2.13)$$

and the operators

$$\hat{D}_n : \mathcal{D}_{G_{n-1}} \bigoplus \mathcal{D}_{Gn-2} \bigoplus \cdots \bigoplus \mathcal{D}_{G_o} \bigoplus \mathcal{H}_1 \bigoplus \mathcal{H}_1 \bigoplus \cdots$$
$$\to \mathcal{D}_{G_n} \bigoplus \mathcal{D}_{G_{n-1}} \bigoplus \cdots \mathcal{D}_{G_0} \bigoplus \mathcal{H}_1 \bigoplus \mathcal{H}_1 \bigoplus \cdots$$
$$D_n = \begin{bmatrix} u_{n+1} D_n u_{n+1} & 0 \\ 0 & I \end{bmatrix}.$$

We can now state the following result which identifies the defect spaces of T_F.

Proposition 2.10 There exist unitary operators identifying $\mathcal{D}_{T_F^*}$ with $\bigoplus_{n=0}^{\infty} \mathcal{D}_{G_n^*}$ and $\mathcal{D}_{T_F}$ with the closure of

$$\mathcal{R}\,(\mathrm{s}-\lim_{n\to\infty}(\hat{D}_o^*\hat{D}_1^*\cdots\hat{D}_n^*\hat{D}_n\cdots\hat{D}_1\hat{D}_o).$$

Proof The proof is based on a realisation of $T_F(T_\infty)$ as a column of infinite length. The identifications of the defect spaces then follow from Proposition A.6.

Using Lemma 2.9

$$\begin{aligned}
&(C_n, C_{n-1}, \cdots, C_o, 0, 0, \cdots) \\
&= ((C_o, \cdots, C_n)u_{n+1}, 0, 0, \cdots) \\
&= (L_n D_{n-1}^n \cdots D_o^n u_{n+1}, 0, 0, \cdots) \\
&= \hat{L}_n \hat{D}_{n-1} \cdots \hat{D}_o,
\end{aligned}$$

where we defined

$$\hat{L}_n = (L_n u_{n+1}, 0, 0, \cdots) : \mathcal{D}_{G_{n-1}} \bigoplus \cdots \bigoplus \mathcal{D}_{G_o} \bigoplus \mathcal{H}_1 \bigoplus \mathcal{H}_1 \bigoplus \cdots \to \mathcal{H}_2.$$

Now,

$$D_{\hat{L}_n}^2 = \hat{D}_n^* \hat{D}_n$$

and there exist unitary operators

$$\omega_n : \mathcal{D}_{G_n} \bigoplus \mathcal{D}_{G_{n-1}} \bigoplus \cdots \bigoplus \mathcal{D}_{G_o} \bigoplus \mathcal{H}_1 \bigoplus \mathcal{H}_1 \bigoplus \cdots \to \mathcal{D}_{\hat{L}_n}$$

such that

$$D_{\hat{L}_n} = \omega_n \hat{D}_n. \tag{2.15)P$$

We begin by defining:

$$\hat{G}_o = \hat{L}_o = (C_o, 0, 0, \cdots) : \ell^2(\mathbf{N}) \bigotimes \mathcal{H}_1 \to \mathcal{H}_2.$$

Then, as

$$(C_1, C_o, 0, \cdots) = \hat{L}_1 \hat{D}_o = \hat{L}_1 \omega_o^* D_{\hat{L}_o} = \hat{L}_1 \omega_o^* D_{\hat{G}_o}$$

we define

$$\hat{G}_1 : \mathcal{D}_{\hat{G}_o} \to \mathcal{H}_2$$

$$\hat{G}_1 = \hat{L}_1\omega_o^*,$$

so that

$$(C_1, C_o, 0, \cdots) = \hat{G}_1 D_{\hat{G}_o}.$$

Moreover, from the relations $D^2_{\hat{G}_1} = \omega_o D^2_{\hat{L}_1}\omega_o^*$, we deduce that $D_{\hat{G}_1} = \omega_o D_{\hat{L}_1}\omega_o^*$ and this shows that $\omega_o^*\mathcal{D}_{\hat{G}_1} = \mathcal{D}_{\hat{L}_1}$. Now suppose we defined the contractions $\hat{G}_o \in \mathcal{L}(\ell^2(\mathbf{N} \bigotimes \mathcal{H}_1, \mathcal{H}_2)$ and $\hat{G}_k \in \mathcal{L}(\mathcal{D}_{\hat{G}_{k-1}}, \mathcal{H}_2)$ for $1 \le k \le n$ such that

$$(C_k, C_{k-1}, \cdots, C_o, 0, \cdots) = \hat{G}_k D_{\hat{G}_{k-1}} \cdots D_{\hat{G}_o} \qquad (2.16)_k$$

and

$$D_{\hat{G}_k} = \omega_o \cdots \omega_{k-1} D_{\hat{L}_k}\omega_{k-1}^* \cdots \omega_o^* \mid \mathcal{D}_{\hat{G}_{k-1}}. \qquad (2.17)_k$$

We find that $\omega_{n-1}^* \cdots \omega_o^* \mathcal{D}_{\hat{G}_n} = \mathcal{D}_{\hat{L}_n}$ and we can define

$$\hat{G}_{n+1} = \hat{L}_{n+1}\omega_n^* \cdots \omega_o^* \mid \mathcal{D}_{\hat{G}_n}$$

in order to get $(2.16)_{n+1}$ and $(2.17)_{n+1}$; these show that

$$T_\infty = \tilde{L}(\{\hat{G}_n\}_{n=0}^\infty).$$

By proposition A.6 we can identify $\mathcal{D}_{T_\infty^*}$ with the closure of $\mathcal{R}$ $(\text{s} - \lim_{n\to\infty} D_{\hat{G}_o} D_{\hat{G}_1} \cdots D^2_{\hat{G}n} \cdots D_{\hat{G}_1} D_{\hat{G}_o})^{\frac{1}{2}}$ and taking (2.15) and (2.17) into account, this space equals the closure of $\mathcal{R}(\text{s} - \lim_{n\to\infty} \hat{D}_o^* \cdots \hat{D}_n^* \hat{D}_n \cdots \hat{D}_o)$. By Proposition A.6 we can identify $\mathcal{D}_{T_\infty}$ with $\bigoplus_{n=0}^{\infty} \mathcal{D}_{\hat{G}_n^*}$. But $\mathcal{D}_{\hat{G}_n^*} = \mathcal{D}_{\hat{L}_n^*} = \mathcal{D}_{L_n^*}$ and since $\mathcal{D}_{L_n^*}$ is identified with $\mathcal{D}_{G_n^*}$, we have the desired result. ∎

In view of Theorem 1.1 we have a characterisation of the finite Blaschke products. By analogy with this characterisation, we call a function $B \in \mathcal{S}(\mathcal{H}_1, \mathcal{H}_2)$ a finite Blaschke product if there exists an $N_o \in \mathbf{N}$ such that its Schur parameter G_{N_o} is an isometry and B is called a finite *–Blaschke product if there exists an $N_o \in \mathbf{N}$ such that G_{N_o} is a co–isometry. We note here only two simple properties of finite Blaschke products. The first one is a consequence of Proposition 2.10 (or (1.42)).

Corollary 2.11 Finite Blaschke products are inner functions and finite *–Blaschke products are *–inner functions.

We can obtain an analogue of Theorem 1.3.

Corollary 2.12 For any $F \in \mathcal{S}(\mathcal{H}_1, \mathcal{H}_2)$, there exists either a sequence of finite Blaschke products or a sequence of finite *–Blaschke products converging to F uniformly on the compact subsets of $\mathbf{D}$.

Proof Let $F \in \mathcal{S}(\mathcal{H}_1, \mathcal{H}_2)$ and let $\{G_n\}_{n=0}^{\infty}$ be its choice sequence. We suppose, of course, that any choice parameter is neither an isometry nor a co–isometry. Now, either there exists a sequence $\{n_k\}_{k\geq 0}$ of positive integers such that $\dim \mathcal{D}_{G_{n_k}} \leq \dim \mathcal{D}_{G^*_{n_k}}$ or there exists a sequence $\{m_k\}_{k\geq 0}$ of positive integers such that $\dim \mathcal{D}_{G^*_{m_k}} \leq \dim \mathcal{D}_{G_{m_k}}$. For the first case we define the finite Blaschke products B_{n_k} given by the choice sequences $\{G_o, G_1, \cdots, G_{n_k-1}, G_{n_k}, V_{n_k}\}$, where $V_{n_k} \in \mathcal{L}(\mathcal{D}_{G_{n_k}}, \mathcal{D}_{G^*_{n_k}})$ is an isometry, and it is clear that $\{B_{n_k}\}_{k\geq 0}$ converges uniformly on the compact subsets of $\mathbf{D}$ to F. In the second case we have to choose V_{n_k} to be co–isometries. ∎

2.3 The structure of positive Toeplitz matrices

In this section we will mainly be concerned with the structure of positive Toeplitz block matrices.

Consider a Hilbert space $\mathcal{H}$ and the operators $S_k \in \mathcal{L}(\mathcal{H})$, $k = 0, \cdots, n$ such that

$$M_n = \begin{bmatrix} S_o & S_1 \cdots & S_n \\ S_1^* & S_o \cdots & S_{n-1} \\ \cdot & & \cdot \\ \cdot & & \cdot \\ \cdot & & \cdot \\ S_n^* & S_{n-1}^* \cdots & S_o \end{bmatrix}$$

is positive. Without loss of generality, we suppose that $S_o = I_{\mathcal{H}}$. For preparing the proof of the main result of this section, we need some new objects. Given a choice sequence $\{G_n\}_{n\geq 1}, G_1 \in \mathcal{L}(\mathcal{H}), G_k \in \mathcal{L}(\mathcal{D}_{G_{k-1}}, \mathcal{D}_{G^*_{k-1}}), k \geq 2$, we define the unitary operators: $U_o = I_{\mathcal{H}}$ and for $n \geq 1$,

$$U_n : \mathcal{H} \bigoplus \bigoplus_{k=1}^{n} \mathcal{D}_{G_k^*} \to \mathcal{H} \bigoplus \bigoplus_{k=1}^{n} \mathcal{D}_{G_k}$$
$$U_n = V(\{G_k\}_{k=1}^{n})(U_{n-1} \bigoplus I_{\mathcal{H}}) \tag{2.18}$$

where, as in the previous sections, we follow the notation in the Appendix. Then we define $E_o = I_{\mathcal{H}}$ and for $n \geq 1$,

$$E_n : \bigoplus_{k=0}^{n} \mathcal{H} \to \mathcal{H} \bigoplus \bigoplus_{k=1}^{n} \mathcal{D}_{G_k}$$

$$E_n = \begin{bmatrix} E_{n-1} & U_{n-1}\tilde{L}(\{G_k\}_{k=1}^n) \\ 0 & D_{G_n} \cdots D_{G_1} \end{bmatrix} \tag{2.19}$$

Lemma 2.13 The following identities hold:

$$E_n = \begin{bmatrix} I & L(\{G_k\}_{k=1}^n)E_{n-1} \\ 0 & D(\{G_k\}_{k=1}^n)E_{n-1} \end{bmatrix} \tag{2.20}$$

and

$$E_n = V(\{G_k\}_{k=1}^n) \begin{bmatrix} L(\{G_k\}_{k=1}^n)^* & E_{n-I} \\ D_{G_n^*} \cdots D_{G_1^*} & 0 \end{bmatrix} \tag{2.21}$$

Proof For simplifying the notation we use in this proof $L_n, \tilde{L}_n, V_n, D_n$ as being the corresponding objects associated with $\{G_k\}_{k=1}^n$. By definition and Lemma A.8, we have

$$\begin{aligned} U_n\tilde{L}_{n+1} &= \begin{bmatrix} L_n & D_{G_1^*} \cdots D_{G_n^*} \\ D_n & -K_n \end{bmatrix} \begin{bmatrix} U_{n-1} & 0 \\ 0 & I \end{bmatrix} \begin{bmatrix} \tilde{L}_n \\ G_{n+1}D_{G_n} \cdots D_{G_1} \end{bmatrix} \\ &= \begin{bmatrix} L_nU_{n-1}\tilde{L}_n + D_{G_1^*} \cdots D_{G_n^*}G_{n+1}D_{G_n} \cdots D_{G_1} \\ D_nU_{n-1}\tilde{L}_n - K_nG_{n+1}D_{G_n} \cdots D_{G_1} \end{bmatrix}. \end{aligned} \tag{2.22)P$$

Now, we prove (2.20) by induction. The step $n = 1$ is obvious. Supposing the formula to be true for n, we get from (2.22) that

$$\begin{aligned} E_{n+1} &= \begin{bmatrix} I & L_nE_{n-1} & L_nU_{n-1}\tilde{L}_n + D_{G_1^*} \cdots D_{G_n^*}G_{n+1}D_{G_n} \cdots D_{G_1} \\ 0 & D_nE_{n-1} & D_nU_{n-1}\tilde{L}_n - K_nG_{n+1}D_{G_n} \cdots D_{G_1} \\ 0 & 0 & D_{G_{n+1}} \cdots D_{G_1} \end{bmatrix} \\ &= \begin{bmatrix} I & (L_n, D_{G_1^*} \cdots D_{G_n^*}G_{n+1}) \begin{bmatrix} E_{n-1} & U_{n-1}\tilde{L}_n \\ 0 & D_{G_n} \cdots D_{G_1} \end{bmatrix} \\ 0 & \begin{bmatrix} D_n & -K_nG_{n+1} \\ 0 & D_{G_{n+1}} \end{bmatrix} \begin{bmatrix} E_{n-1} & U_{n-1}L_n \\ 0 & D_{G_n} \cdots D_{G_1} \end{bmatrix} \end{bmatrix} \\ &= \begin{bmatrix} I & L_{n+1}E_n \\ 0 & D_{n+1}E_n \end{bmatrix} \end{aligned}$$

and the first relation in Lemma (2.13) is proved. For the second one we use Proposition A.9

$$V_n \begin{bmatrix} L_n^* & E_{n-1} \\ D_{G_n^*} \cdots D_{G_1^*} & 0 \end{bmatrix} = \begin{bmatrix} L_n & D_{G_1^*} \cdots D_{G_n^*} \\ D_n & -K_n \end{bmatrix} \begin{bmatrix} L_n^* & E_{n-1} \\ D_{G_n^*} \cdots D_{G_1^*} & 0 \end{bmatrix}$$

$$= \begin{bmatrix} L_nL_n^* + D_{G_1^*}\cdots D^2_{G_n^*}\cdots D_{G_1^*} & L_nE_{n-1} \\ D_nL_n^* - K_nD_{G_n^*}\cdots D_{G_1^*} & D_nE_{n-1} \end{bmatrix}$$
$$= \begin{bmatrix} I & L_nE_{n-1} \\ O & D_nE_{n-1} \end{bmatrix}$$
$$= E_n.$$

■

Now we can prove the main result of this section.

Theorem 2.14 There exists a one–to–one correspondence between the set of positive Toeplitz block matrices M_n and the set of families of contractions $\{G_k\}_{k=1}^n, G_1 \in \mathcal{L}(\mathcal{H}), G_k \in \mathcal{L}(\mathcal{D}_{G_{k-1}}, \mathcal{D}_{G^*_{k-1}}), k \geq 2$, expressed by the formulas

$$S_1 = G_1$$
$$S_k = L(\{G_p\}_{p=1}^{k-1})U_{k-2}\tilde{L}(\{G_p\}_{p=1}^{k-1}) + D_{G_1^*}\cdots D_{G^*_{k-1}}G_kD_{G_{k-1}}\cdots D_{G_1}$$

for $k \geq 1$.

Proof We can prove this result by induction on n. We use the same convention regarding the notation of $L\{G_p\}_{p=1}^k), \cdots$ as in the proof of Lemma 2.13. By Theorem A.1 we deduce that $M_1 \geq 0$ if and only if $S_1 = G_1$ is a contraction of $\mathcal{H}$. Furthermore, we find the contractions$\{G_k\}_{k=2}^n$ by proving the following statements for $n \geq 1$

$$(S_1, \cdots, S_n) = L_nE_{n-1}, \qquad (2.23)_n$$
$$(S_n, \cdots, S_1)^t = E^*_{n-1}U_{n-1}\tilde{L}_n, \qquad (2.24)_n$$
$$M_n = E_n^*E_n \qquad (2.25)_n$$

and there exists a uniquely determined contraction

$$G_{n+1} : \mathcal{D}_{G_n} \to \mathcal{D}_{G_n^*} \quad \text{with}$$
$$S_{n+1} = L_nU_{n-1}\tilde{L}_n + D_{G_1^*}\cdots D_{G_n^*}G_{n+1}D_{G_n}\cdots D_{G_1}. \qquad (2.26)$$

For $n = 1$, the first three statements are obviously true. In order to prove $(2.26)_1$, we write

$$M_2 = \begin{bmatrix} I & (S_1, S_2) \\ \begin{bmatrix} S_1^* \\ S_2^* \end{bmatrix} & \begin{bmatrix} I & S_1 \\ S_1^* & I \end{bmatrix} \end{bmatrix}$$

and, according to Theorem A.1, $M_2 \geq 0$ if and only if

$$\begin{bmatrix} I & S_1 \\ S_1^* & I \end{bmatrix} \geq \begin{bmatrix} S_1^* \\ S_2^* \end{bmatrix} (S_1, S_2) \quad \text{or} \quad E_1^* E_1 \geq (S_1, S_2)^*(S_1, S_2).$$

By Proposition A.2, there exists a contraction

$$G' = (G_1', G_2')$$

such that

$$(S_1, S_2) = (G_1', G_2') \begin{bmatrix} I & G_1 \\ 0 & D_{G_1} \end{bmatrix} = (G_1', G_1' G_1 + G_2' D_{G_1})$$

Consequently, $G_1' = G_1$ and, by Theorem A.4, $G_2' = D_{G_1} G_2$ with a uniquely determined contraction $G_2 : \mathcal{D}_{G_1} \to \mathcal{D}_{G_1^*}$, and it follows that $S_2 = S_1^2 + D_{G_1^*} G_2 D_{G_1}$. For proving $(2.23)_n$, we have

$$\begin{aligned}(S_1, \cdots, S_n) &= (L_{n-1} E_{n-2}, L_{n-1} U_{n-2} \tilde{L}_{n-1} + D_{G_1^*} \cdots D_{G_{n-1}^*} G_n D_{G_{n-1}} \cdots D_{G_1}) \\ &= L_n \begin{bmatrix} E_{n-2} & U_{n-2} \tilde{L}_{n-1} \\ 0 & D_{G_{n-1}} \cdots D_{G_1} \end{bmatrix} = L_n E_{n-1}.\end{aligned}$$

For $(2.24)_n$, we have

$$\begin{aligned}(S_n, \cdots, S_1)^{\mathrm{t}} &= (L_{n-1} U_{n-2} \tilde{L}_{n-1} + D_{G_1^*} \cdots D_{G_{n-1}^*} G_n D_{G_{n-1}} \cdots D_{G_1}, E_{n-2}^* U_{n-2} \tilde{L}_{n-1})^{\mathrm{t}} \\ &= \begin{bmatrix} E_{n-1} & D_{G_1^*} \cdots D_{G_n^*} \\ E_{n-2}^* & 0 \end{bmatrix} V_{n-1}^* U_{n-1} \tilde{L}_n.\end{aligned}$$

But by Lemma 2.13. the last term is exactly $E_{n-1}^* U_{n-1} \tilde{L}_n$. We can deduce $(2.25)_n$ from

$$M_n = \begin{bmatrix} M_{n-1} & (S_n, \cdots, S_1)^{\mathrm{t}} \\ (S_n^*, \cdots, S_1^*) & I \end{bmatrix} = \begin{bmatrix} E_{n-1}^* E_{n-I} & E_{n-1}^* U_{n-1} \tilde{L}_n \\ \tilde{L}_n^* U_{n-1}^* E_{n-1} & I \end{bmatrix} = E_n^* E_n.$$

Finally, by Theorem A.1, $M_{n+1} \geq 0$ if and only if

$$M_n \geq (S_1, \cdots, S_{n+1})^*(S_1, \cdots, S_{n+1})$$

or

$$E_n^* E_n \geq (S_1, \cdots, S_{n+1})^*(S_1, \cdots, S_{n+1}).$$

Then, there exists a contraction $G' = (G'_1, \cdots, G'_n, G'_{n+1})$, which, by $(2.23)_n$ is of the form (L_n, G'_{n+1}) and by Theorem A.4, $G'_{n+1} = D_{G_1^*} \cdots D_{G_n^*} G_{n+1}$ with a uniquely determined contraction $G_{n+1} : \mathcal{D}_{G_n} \to \mathcal{D}_{G_n^*}$. Moreover,

$$S_{n+1} = L_n U_{n-1} \tilde{L}_n + D_{G_n^*} \cdots D_{G_n^*} G_{n+1} D_{G_n} \cdots D_{G_1}$$

which is exactly $(2.26)_n$, and the proof is finished. ∎

Remark 2.15 We also obtained in the above proof the factorisation

$$M_n = E_n^* E_n$$

with E_n an upper triangular operator. Such a factorisation is usually called Cholesky factorisation and one of its consequences is the following determinant formula for M_n : if S_k are $r \times r$ matrices, $r \in \mathbf{N}$, then

$$\det M_n = \prod_{k=1}^{n} (\det D_{G_k})^{2(n-k+1)}. \tag{2.27}$$

Remark 2.16 When S_o is not the identity operator, G_1 is a contraction on $\mathcal{L}(\overline{\mathcal{R}(S_o)})$ and

$$S_1 = S_o^{\frac{1}{2}} G_1 S_o^{\frac{1}{2}}.$$

$$S_n = S_o^{\frac{1}{2}} (L(\{G_k\}_{k=1}^n U_{n-2} \tilde{L}(\{G_k\}_{k=1}^n) + D_{G_1^*} \cdots D_{G_{n-1}^*} G_n D_{G_{n-1}} \cdots D_{G_1}) S^{\frac{1}{2}}$$

with uniquely determined contractions $G_k : \mathcal{D}_{G_{k-1}} \to \mathcal{D}_{G_{k-1}^*}$, $k = 1, \cdots, n$.

Supposing again that $\dim \mathcal{H} < \infty$, the identity (2.27) becomes

$$\det M_n = (\det S_o)^{n+1} \prod_{k=1}^{n} (\det D_{G_k})^{2(n-k+1)}. \tag{2.28}$$

Remark 2.17 Consider a positive definite kernel on the set of integers, i.e. an application

$$\mathcal{T} : \mathbf{Z} \times \mathbf{Z} \to \mathcal{L}(\mathcal{H})$$

such that for each choice of vectors $h_1, \cdots, h_n$ in $\mathcal{H}$ and for any $k_1, \cdots, k_n$ in $\mathbf{Z}$, the following inequality

$$\sum_{i,j=1}^{n} (\mathcal{T}(k_1, k_j) h_j, h_i) \geq 0$$

holds. We also suppose that the kernel is Toeplitz, i.e. $\mathcal{T}(k_i, k_j)$ depends only on the difference $k_j - k_i$ of the indices. In this case $\mathcal{T}$ will be a positive definite kernel if and only if the matrices M_n built up on the coefficients $S_n = \mathcal{T}(i, i+n)$ of the kernel are positive for $n \geq 0$. In view of Theorem 2.14, we obtain a one–to–one correspondence between the positive definite Toeplitz kernels on $\mathbf{Z}$ (with $S_o = I_{\mathcal{H}}$) and the sequences of contractions $\{G_n\}_{n=1}^{\infty}, G_1 \in \mathcal{L}(\mathcal{H}), G_n \in \mathcal{L}(\mathcal{D}_{G_{n-1}}, \mathcal{D}_{G_{n-1}^*}), n \geq 2$. In this way, with any positive definite Toeplitz kernel on $\mathbf{Z}$ we have associated a uniquely determined choice sequence $\{G_n\}_{n=1}^{\infty}$.

Take $\mathcal{T}$ as a positive definite Toeplitz kernel on $\mathbf{Z}$ given by the choice sequence $\{G_n\}_{n=1}^{\infty}$; define $G_o = 0_{\mathcal{H}}$, so that, $\{G_n\}_{n=0}^{\infty}$ remains a choice sequence. With this choice sequence we associate by Theorem 2.1 a contraction T_{∞}. We now have to explain the connection between $\mathcal{T}$ and T_{∞}.

Proposition 2.18 Let $\{G_n\}_{n=1}^{\infty}$ be a choice sequence ($G_1 \in \mathcal{L}(\mathcal{H})$) which is completed with $G_0 = 0_{\mathcal{H}}$. Let $\mathcal{T}$ be the positive definite Toeplitz kernel associated with $\{G_n\}_{n=1}^{\infty}$ by Theorem 2.14 and let T_{∞} be the analytic Toeplitz contraction associated with $\{G_n\}_{n=0}^{\infty}$ by Theorem 2.1. Then:

$$C_0 = 0$$

$$C_1 = G_1 = S_1$$

$$C_n S_n - \sum_{k=1}^{n-1} C_k S_{n-k}, \quad n > 1.$$

Proof The relation in the statement of the Proposition is equivalent to

$$S_n = (C_1, \cdots, C_{n-1})(S_{n-1}, \cdots, S_1)^{\mathrm{t}} + C_n. \tag{2.29}$$

It is now a simple observation that all formulas in the proof of Theorem 2.1 still hold if we replace the choice sequence $\{G_n\}_{n=0}^{\infty}$ with $\{G_n\}_{n=1}^{\infty}$ and we start with the elements indexed by 1. Then, by $(2.1)_{n-1}$,

$$(C_1, \cdots, C_{n-1}) = L(\{G_k\}_{k=1}^{n-1})\Omega(\{G_k\}_{k=1}^{n-2})D_{B\{G_k\}_{k=1}^{n-2}}.$$

Moreover, by $(2.24)_{n-1}$,

$$(S_{n-1}, \cdots, S_1)^{\mathrm{t}} = E_{n-2}^* U_{n-2} \tilde{L}(\{G_k\}_{k=1}^{n-1}).$$

Having also the expressions of S_n and C_n in terms of the choice sequence, it remains to prove that

$$U_n = \Omega(\{G_k\}_{k=1}^n) D_{B(\{G_k\}_{k=1}^n)} E_N^* U_n + Q(\{G_k\}_{k=1}^n), \quad n \in \mathbf{N},$$

but this can easily be done by induction. ∎

Remark 2.19 The relations in Proposition 2.18 remind us of the relations (1.16) between the coefficients of functions in the Schur class and the functions in the Carathéodory class. Also, in the operatorial case, we can consider the Carathéodory class $\mathcal{C}(\mathcal{H})$ of analytic functions in $\mathbf{D}$ with positive real part and taking values in $\mathcal{L}(\mathcal{H})$, for a separable Hilbert space $\mathcal{H}$. The relations in Proposition 2.18 reflect the correspondence

$$F(z) = (G(z) - I)(G(z) + I)^{-1}$$

which is one to one between the functions $F \in \mathcal{S}(\mathcal{H})$ with $F(0) = 0$ and the functions $G \in \mathcal{C}(\mathcal{H})$ with $G(0) = I$, when we take into consideration their Taylor series

$$F(z) = zC_1 + z^2 C_2 + \cdots$$

and

$$G(z) = I + 2zS_1 + 2z^2 S_2 + \cdots$$

∎

2.4 Naimark dilations

Let $\mathcal{H}$ be a separable Hilbert space. An $\mathcal{L}(\mathcal{H})$–valued semispectral measure μ on $\mathbf{T}$ is a linear positive map from $C(\mathbf{T})$, the set of continuous functions on the unit circle, into $\mathcal{L}(\mathcal{H})$.

We will usually suppose that $\mu(1) = I_{\mathcal{H}}$. The Fourier coefficients of μ are $S_n = \mu(\chi_{-n})$, where $\chi_n(e^{it}) = e^{int}, n \in \mathbf{Z}$.

Proposition 2.20 If $\{S_n\}_{n\in\mathbf{Z}}$ are the Fourier coefficients of a semispectral measure on $\mathbf{T}$, then the Toeplitz kernel $\mathcal{T}(k, k+n) = S_n$, $k, n \in \mathbf{Z}$, is positive definite.

Proof This proposition is mainly a restatement of the fact that a linear positive map acting between two C^*-algebras, the first one being commutative, is completely positive. More precisely, the applications

$$\mu_n((f_{ij})_{1,j=1}^n) = (\mu(f_{ij}))_{i,j=1}^n, \quad f_{ij} \in C(\mathbf{T})$$

are positive for $n \in \mathbf{N}$. As the matrices

$$\begin{bmatrix} \chi_o & \chi_{-1} & \cdots & \chi_n \\ \chi_1 & \chi_o & \cdots & \chi_{-n+1} \\ \cdot & & & \cdot \\ \cdot & & & \cdot \\ \cdot & & & \cdot \\ \chi_n & \cdots & & \chi_o \end{bmatrix}$$

are obviously positive, it follows that

$$M_n = \begin{bmatrix} S_o & S_1 & \cdots & S_n \\ S_1^* & S_o & \cdots & S_{n-1} \\ \cdot & & & \cdot \\ \cdot & & & \cdot \\ \cdot & & & \cdot \\ S_n^* & & \cdots & S_o \end{bmatrix}$$

are positive operators for $n \geq 0$.

In view of Proposition 2.20, we can naturally use the results in the previous section for the Fourier coefficients of a semispectral measure on $\mathbf{T}$. As the density of trigonometric polynomials in $C(\mathbf{T})$ asssures us that μ is uniquely determined by its Fourier coefficients, to any μ we have associated a choice sequence.

The main purpose of this section is to give an explicit construction of the so–called Naimark dilation of μ in terms of its choice sequence. By a Naimark dilation we mean here a pair $\{\mathcal{K}, W\}$ formed by a Hilbert space $\mathcal{K}$ containing $\mathcal{H}$ and by a unitary operator W in $\mathcal{L}(\mathcal{K})$ such that

$$S_n = P_{\mathcal{H}}^{\mathcal{K}} W^n \mid \mathcal{K}, \quad n \in \mathbf{Z}. \tag{2.30}$$

($P_{\mathcal{H}}^{\mathcal{K}}$ denotes the orthogonal projection of $\mathcal{K}$ onto $\mathcal{H}$).

In the supplementary condition of minimality:

$$\mathcal{K} = \bigvee_{n=-\infty}^{\infty} W^n \mathcal{H} \tag{2.31}$$

the Naimark dilation is unique in the following sense: if $\{\mathcal{K}_1, W_1\}$ and $\{\mathcal{K}_2, W_2\}$ are minimal Naimark dilations of μ then there exists a unitary operator $\omega : \mathcal{K}_1 \to \mathcal{K}_2$ such that $\omega \mid \mathcal{H} = I_{\mathcal{H}}$ and $\omega W_1 = W_2 \omega$. To see this, it is sufficient to define

$$\sum_{i=-N_1}^{N_2} W_1^i h_i \overset{\omega}{\to} \sum_{i=-N_1}^{N_2} W_2^i h_i. \tag{2.32}$$

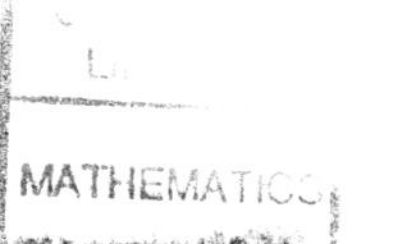

This is correctly defined, isometric, and in view of minimality, it extends to a unitary operator between $\mathcal{K}_1$ and $\mathcal{K}_2$.

Now, we fix a semispectral measure μ on $\mathbf{T}$ and $\{S_n\}_{n\in\mathbf{Z}}$ are its Fourier coefficients. Accordingly to Proposition 2.20 and Remark 2.17, we consider $\{G_n\}_{n=1}^{\infty}$, the choice sequence associated with $\{S_n\}_{n\in\mathbf{Z}}$ and, therefore, with μ. Define the row contraction $L = L(\{G_n\}_{n=1}^{\infty}) : \bigoplus_{n=1}^{\infty} \mathcal{D}_{G_n} \to \mathcal{H}$ (we maintain the convention that $G_o = 0_{\mathcal{H}}$). We can take into account the unitary operator:

$$
\begin{aligned}
&W_{red} : \mathcal{D}_*(L) \bigoplus \bigoplus_{n=0}^{\infty} \mathcal{D}_{G_n} \to \bigoplus_{n=0}^{\infty} \mathcal{D}_{G_n} \\
&W_{red} = \begin{bmatrix} I & 0 \\ 0 & \alpha(L) \end{bmatrix} J(L) \begin{bmatrix} 0 & I \\ \beta^*(L) & 0 \end{bmatrix}
\end{aligned}
\tag{2.33}
$$

and then we define the space

$$
\mathcal{K} = \cdots \bigoplus \mathcal{D}_*(L) \bigoplus \mathcal{D}_*(L) \bigoplus \mathcal{H} \bigoplus \bigoplus_{n=1}^{\infty} \mathcal{D}_{G_n}
\tag{2.34}
$$

and the unitary operator

$$
\begin{aligned}
&W : \mathcal{K} \to \mathcal{K} \\
&W = I \bigoplus W_{red}
\end{aligned}
\tag{2.35}
$$

which is written with respect to the decompositions:

$$
\mathcal{K} = (\cdots \bigoplus \mathcal{D}_*(L)) \bigoplus \mathcal{D}_*(L) \bigoplus \bigoplus_{n=0}^{\infty} \mathcal{D}_{G_n})
$$

and

$$
\mathcal{K} = (\cdots \bigoplus \mathcal{D}_*(L)) \bigoplus (\bigoplus_{n=0}^{\infty} \mathcal{D}_{G_n}).
$$

If we denote $\mathcal{K}_+ = \bigoplus_{n=0}^{\infty} \mathcal{D}_{G_n}$ we can write W in the following matrix form

$$W = \begin{bmatrix} & \cdot & \cdot & & \\ & \cdot & \cdot & & \\ & \cdot & \cdot & & \\ \cdots & I & 0 & 0 & \\ & & & & 0 \\ & \cdots & I & 0 & \\ & \cdots & 0 & H_\infty(L) & \\ & \cdots & 0 & -Z_1 & \\ & \cdots & 0 & -Z_2 & W \mid \mathcal{K}_+ \\ & & \cdot & & \\ & & \cdot & & \\ & & \cdot & & \end{bmatrix}$$

where

$$H_\infty(L) = (\mathrm{s} - \lim_{n\to\infty} D_{G_1^*} \cdots D^2_{G_n^*} \cdots D_{G_1^*})^{\frac{1}{2}},$$
$$Z_k : \mathcal{D}_*(L) \to \mathcal{D}_{G_k}. \quad k \geq 1$$

are the operators described by Lemma A.5 and Lemma A.10.

The operator

$$W_+ : \mathcal{K}_+ \to \mathcal{K}_+$$
$$W_+ = W \mid \mathcal{K}_+$$

is an isometry,

$$W_+ = \begin{bmatrix} L \\ D_\infty(L) \end{bmatrix}$$

where $D_\infty(L)$ is given by Lemma A.5 and has the matrix

$$D_\infty(L) = \begin{bmatrix} D_{G_1} & -G_1^*G_2 & -G_1 D_{G_2} G_3 & \cdots \\ 0 & D_{G_2} & -G_2^*G_3 & \cdots \\ 0 & 0 & D_{G_3} & \cdots \\ \cdot & \cdot & & \\ \cdot & \cdot & & \\ \cdot & \cdot & & \end{bmatrix}.$$

We can now prepare the proof of the main result of this section.

Define the spaces

$$\mathcal{K}_n = \bigoplus_{k=0}^{n-1} \mathcal{D}_{G_k} \tag{2.36}$$

and the operators

$$W_n : \mathcal{K}_n \to \mathcal{K}_n$$
$$W_n = V(\{G_k\}_{k=1}^{n-1})(I \bigoplus G_n). \tag{2.37}$$

Lemma 2.21 $W_+ = \mathrm{s} - \lim\limits_{n\to\infty} W_n P_{\mathcal{K}_n}^{\mathcal{K}}$.

Proof We know from the Appendix that

$$V(\{G_k\}_{k=1}^{n-1}) = \begin{bmatrix} L(\{G_k\}_{k=1}^{n-1} & D_{G_1^*} \cdots D_{G_{n-1}^*} \\ D(\{G_k\}_{k=1}^{n-1}) & -K(\{G_k\}_{k=1}^{n-1}) \end{bmatrix}.$$

This implies

$$W_n = \begin{bmatrix} W_{n-1} & & D_{G_1^*} \cdots D_{G_{n-1}^*} G_n \\ 0 & \cdots 0 \ D_{G_{n-1}} & -K(\{G_k\}_{k=1}^{n-1}) G_n \end{bmatrix} \tag{2.38}$$

which shows that $\{W_n P_{\mathcal{K}_n}^{\mathcal{K}} f\}_{n=1}^{\infty}$ is a Cauchy sequence for every $f \in \bigcup\limits_{n=1}^{\infty} \mathcal{K}_n$. As $\| W_n P_{\mathcal{K}_n}^{\mathcal{K}} \| \leq 1$, it follows that there exists the strong operatorial limit $\mathrm{s}\lim\limits_{n\to\infty} W_n P_{\mathcal{K}_n}^{\mathcal{K}}$. Then, by the definition of W_+

$$P_{\mathcal{K}_n}^{\mathcal{K}} W_+ \mid \mathcal{K}_n = W_n, \quad n \geq 1, \tag{2.39}$$

and this shows that $\mathrm{s} - \lim\limits_{m\to\infty} W_n P_{\mathcal{K}_n}^{\mathcal{K}} = W_+$. ∎

Lemma 2.22 For every $n \geq 1$, we have the equality

$$P_{\mathcal{H}}^{\mathcal{K}} W_+^n \mid \mathcal{H} = P_{\mathcal{H}}^{\mathcal{K}_n} W_n^n \mid \mathcal{H}.$$

Proof From the equality

$$P_{\mathcal{K}_{n+1}}^{\mathcal{K}} W_+ \mid \mathcal{K}_n = W_+ \mid \mathcal{K}_n, \quad n \geq 1 \tag{2.40}$$

it follows that

$$P_{\mathcal{K}_{n+k}}^{\mathcal{K}} W_+^k \mid \mathcal{K}_n = W_+^k \mid \mathcal{K}_n, \quad n, k \geq 1 \tag{2.41}$$

and now

$$\begin{aligned} P_{\mathcal{H}}^{\mathcal{K}} W_n^n \mid \mathcal{H} &= P_{\mathcal{H}}^{\mathcal{K}}(P_{\mathcal{K}_n}^{\mathcal{K}} W_+ P_{\mathcal{K}_n}^{\mathcal{K}}) \cdots (P_{\mathcal{K}_n}^{\mathcal{K}} W_+ P_{\mathcal{K}_n}^{\mathcal{K}}) \mid \mathcal{H} \\ &= P_{\mathcal{H}}^{\mathcal{K}}(P_{\mathcal{K}_n}^{\mathcal{K}} W_+) \cdots (P_{\mathcal{K}_n}^{\mathcal{K}} W_+) P_{\mathcal{K}_n}^{\mathcal{K}} W_+ \mid \mathcal{H} = \cdots \\ &= P_{\mathcal{H}}^{\mathcal{K}} W_+^n \mid \mathcal{H}. \end{aligned}$$

Now we can obtain the main result of this section.

Theorem 2.23 The pair $\{\mathcal{K}, W\}$ given by (2.34) and (2.35) is a minimal Naimark dilation of the semispectral measure μ.

Proof In view of Lemmas 2.21 and 2.22, we have to prove that

$$S_n = P_{\mathcal{H}}^{\mathcal{K}} W_n^n \mid \mathcal{H}, \quad n \geq 1.$$

First of all, it is clear from the proof of Lemma 2.21 that

$$(I, 0_{n-1}) W_n = L(\{G_k\}_{k=1}^n). \tag{2.42}$$

Then we prove by induction that

$$W_n^{n-1} \begin{bmatrix} I \\ 0_{n-1} \end{bmatrix} = \begin{bmatrix} U_{n-2} \tilde{L}(\{G_k\}_{k=1}^{n-1}) \\ D_{G_{n-1}} \cdots D_{G_1} \end{bmatrix}, \quad n \geq 2. \tag{2.43}$$

For $n = 2$, this is obvious. Then,

$$\begin{aligned} W_{n+1}^n \begin{bmatrix} I \\ 0_n \end{bmatrix} &= \begin{bmatrix} W_n & * \\ 0, \cdots, D_{G_n} & * \end{bmatrix}^n \begin{bmatrix} I \\ 0_n \end{bmatrix} \\ &= \begin{bmatrix} W_n & * \\ 0, \cdots, D_{G_n} & * \end{bmatrix} \begin{bmatrix} W_n^{n-1} \begin{bmatrix} I \\ 0_{n-1} \end{bmatrix} & 0 \\ 0 & 0 \end{bmatrix} \\ &= \begin{bmatrix} W_n \begin{bmatrix} U_{n-2} \tilde{L}(\{G_k\}_{k=1}^{n-1}) \\ D_{G_{n-1}} \cdots D_{G_1} \end{bmatrix} \\ D_{G_n} \cdots D_{G_1} \end{bmatrix}. \end{aligned}$$

But now,

$$\begin{aligned} W_n \begin{bmatrix} U_{n-2} \tilde{L}\{G_k\}_{k=1}^{n-1}) \\ D_{G_{n-1}} \cdots D_{G_1} \end{bmatrix} &= V_{n-1} \begin{bmatrix} I_{n-1} & 0 \\ 0 & G_n \end{bmatrix} \begin{bmatrix} U_{n-2} & 0 \\ 0 & I \end{bmatrix} \begin{bmatrix} \tilde{L}(\{G_k\}_{k=1}^{n-1}) \\ D_{G_{n-1}} \cdots D_{G_1} \end{bmatrix} \\ &= V_{n-1} \begin{bmatrix} U_{n-2} & 0 \\ 0 & I \end{bmatrix} \begin{bmatrix} \tilde{L}(\{G_k\}_{k=1}^{n-1}) \\ G_n D_{G_{n-1}} \cdots D_{G_1} \end{bmatrix} \\ &= U_{n-1} \tilde{L}(\{G_k\}_{k=1}^n). \end{aligned}$$

Now, in order to conclude the proof, we put (2.42) and (2.43) together and we get:

$$P_{\mathcal{H}}^{\mathcal{K}} W_n^n \mid \mathcal{H} = (I, 0_{n-1}) W_n W_n^{n-1} \begin{bmatrix} I \\ O_{n-1} \end{bmatrix}$$

$$= L(\{G_k\}_{k=1}^{n-1})U_{n-2}\tilde{L}(\{G_k\}_{k=1}^{n-1}) + D_{G_1^*}\cdots D_{G_{n-1}^*}G_nD_{G_{n-1}}\cdots D_{G_1}$$

and by Theorem 2.14, we have

$$P_{\mathcal{H}}^{\mathcal{K}}W^n \mid \mathcal{H} = P_{\mathcal{H}}^{\mathcal{K}}W_+^n \mid \mathcal{H} = P_{\mathcal{H}}^{\mathcal{K}_n}W_n^n \mid \mathcal{H} = S_n, \quad n \geq 1.$$

Regarding the minimality of $\{\mathcal{K}, W\}$, we take $f = (h_o, h_1, \cdots) \in \mathcal{K}_+, f \perp W^n\mathcal{H}, n \in \mathbf{N}$. As

$$W_+^n h = (\cdots, *, *, D_{G_n}\cdots D_{G_1}h, 0, 0, \cdots), \quad h \in \mathcal{H}$$

we obtain $h_n = 0, n \in \mathbf{N}$, so that $\mathcal{K}_+ = \bigvee_{n=0}^{\infty} W_+^n\mathcal{H}$. Furthermore, we take $f \in \mathcal{K}, f \perp W^n\mathcal{H}, n \in \mathbf{Z}$, of the form $f = (\cdots, h_{-1}, h_o, h_1, \cdots)$ and we obtain $h_n = 0, n \geq 0$. As for $n > 0$,

$$W^{*n}h = (\cdots, 0, H_\infty(L)h, *, *, \cdots),$$

it also follows that $h_n = 0$, for $n < 0$, so that $\mathcal{K} = \bigvee_{n=-\infty}^{\infty} W^n\mathcal{H}$ and this means that $\{\mathcal{K}, W\}$ is a minimal Naimark dilation of μ. ∎

In view of the existence of the operator (2.32) we will refer to $\{\mathcal{K}, W\}$ given by (2.34) and (2.35) as the Naimark dilation of μ.

Remark 2.24 If we consider a positive Toeplitz kernel on $\mathbf{Z}$, then if $\{G_n\}_{n=1}^{\infty}$ is its choice sequence, we can take into account the pair $\{\mathcal{K}, W\}$ given by (2.34) and (2.35). Then we define the semispectral measure

$$\begin{aligned} &\mu : C(\mathbf{T}) \to \mathcal{L}(\mathcal{H}) \\ &\mu(f) = P_{\mathcal{H}}^{\mathcal{K}} f(W) \mid \mathcal{H} \end{aligned} \qquad (2.44)P$$

($f(W)$ is the functional calculus of W) and this semispectral measure has as Fourier coefficients the coefficients of the Toeplitz kernel we started with. This is the converse of Proposition 2.20. ∎

Remark 2.25 Consider a contraction $T \in \mathcal{L}(\mathcal{H})$ and define the choice sequence $G_1 = T, G_k = 0, k > 1$. Then the kernel associated by Theorem 2.14 is

$$S_n = \begin{cases} T^n & n > 0 \\ T & n = 0 \\ T^{*|n|} & n < 0 \end{cases}$$

and if we now take its Naimark dilation, we get

$$\mathcal{K} = \cdots \bigoplus \mathcal{D}_{T^*} \bigoplus \mathcal{D}_{T^*} \bigoplus \mathcal{H} \bigoplus \mathcal{D}_T \bigoplus \mathcal{D}_T \bigoplus \cdots$$

$$W = \begin{bmatrix} & \cdot & \cdot & & & \\ & \cdot & \cdot & & & \\ & \cdot & \cdot & & & \\ \cdots & I & 0 & & & \\ \cdots & 0 & I & 0 & 0 & \cdots \\ \cdots & 0 & D_{T^*} & T & 0 & \cdots \\ \cdots & & -T^* & D_T & 0 & \cdots \\ \cdots & & & 0 & I & \cdots \\ & & & \cdot & \cdot & \\ & & & \cdot & \cdot & \\ & & & \cdot & \cdot & \end{bmatrix}$$

which is the Schäffer form of the $Sz.$ Nagy dilation of the contraction T.

Remark 2.26 Taking a positive kernel on $\mathbf{Z}$ with values in $\mathcal{L}(\mathcal{H})$, denoted by $\mathcal{T}$, we can make the following construction: on the set of all sequences in $\bigoplus_{n\in\mathbf{Z}} \mathcal{H}$ with finite support, we define the inner product

$$(f,g)_{\mathcal{T}} = (M_n(\mathcal{T})f,g)$$

for a sufficiently large n. (We denote by $\{S_n\}_{n\in\mathbf{Z}}$ the Fourier coefficients of $\mathcal{T}$, and $M_n(\mathcal{T})$ are the matrices built up on these coefficients.) This inner product is positive; factoring out its kernel if necessary and completing the resulting space, we obtain a Hilbert space denoted by $\ell^2(\mathbf{Z},\mathcal{T})$. Based on Theorem 2.23, we can obtain the following identification of this space:

$$\Omega : \ell^2\mathbf{Z},\mathcal{T}) \to \mathcal{K} \tag{2.45}$$

where for an element f of finite support in $\bigoplus_{n\in\mathbf{Z}} \mathcal{H}$ we define

$$\Omega\hat{f} = (\cdots, W^* \mid \mathcal{H}, \boxed{P_{\mathcal{H}}^{\mathcal{K}} \mid \mathcal{H}}, W \mid \mathcal{H}, W^2 \mid \mathcal{H}, \cdots)f;$$

$\hat{f}$ is the class of f in $\ell^2(\mathbf{Z},\mathcal{T})$ and $\boxed{}$ denotes the zero position. In view of (2.43) the Cholesky factorisation of $M_n(\mathcal{T})$ is also given by

$$E_n = P(_{\mathcal{H}}^{\mathcal{K}_{n+1}} \mid \mathcal{H}, W_{n+1} \mid \mathcal{H}, \cdots, W_{n+1}^n \mid \mathcal{H}) \tag{2.46}$$

and this shows that Ω is an isometry. Then, Ω extends to $\ell^2(\mathbf{Z},\mathcal{T})$ and in view of the minimality of W, this extension, also denoted by Ω, is a unitary operator. ∎

Remark 2.27 On $\ell^2(\mathbf{Z},\mathcal{T})$ we can define the shift operator

$$S_+ : \ell^2(\mathbf{Z},\mathcal{T}) \to \ell^2(\mathbf{Z},\mathcal{T})$$
$$S_+(\cdots,h_{-1},\boxed{h_o},h_1,\cdots) = (\cdots h_{-2},\boxed{h_{-1}},h_o,\cdots) \tag{2.47}$$

first for elements with finite support and then we can extend it to the whole $\ell^2(\mathbf{Z},\mathcal{T})$ because it is bounded (actually S_+ is a unitary operator in view of the fact that $\mathcal{T}$ is Toeplitz). The following relation obviously holds:

$$W\Omega = \Omega S_+. \tag{2.48}$$

We return now to the trigonometric moment problem. First, from Theorem 2.14 we deduce a simple solution of the following operatorial version.

'Given the operators $S_k \in \mathcal{L}(\mathcal{H})$, $k = 0,\cdots,n$ ($S_o = 1$), it is required to find conditions for the existence of a semispectral measure μ on the unit circle such that $S_k = \mu(\chi_{-k})$, $k = 0,\cdots,n$.'

Theorem 2.28 The operatorial trigonometric moment problem can be solved if and only if the block matrix

$$M_n = \begin{bmatrix} I & S_1 & \cdots & S_n \\ S_1^* & I & \cdots & S_{n-1} \\ \cdot & & & \\ \cdot & & & \\ \cdot & & & \\ S_n^* & & \cdots & I \end{bmatrix}$$

built up on the given operators $\{S_k\}_{k=1}^n$ is positive.

Proof By proposition 2.20 the condition is necessary. Then, supposing $M_n \geq 0$, a family $\{G_k\}_{k=1}^n$ of choice parameters is uniquely associated with M_n by Theorem 2.14. In view of the same Theorem 2.14 and of Remark 2.24, any continuation of this family to a choice sequence of infinite length will produce a solution of the problem. ∎

Let us continue with a related extremal problem. Take $\mathcal{H}$ to be of finite dimension and let us consider the following maximum determinant problem:

'Given the matrices S_k, $k = 0,\cdots,n$ ($S_o = I$) such that the trigonometric moment problem with these data can be solved, and given a number $N > n$, it is required to compute

$m(\{S_k\}_{k=0}^n) = \max \{ \det M_N(\{S_k\}_{k=0}^N) \mid \{S_p\}_{p=n+1}^N$ are such that $M_N(\{S_k\}_{k=0}^N) \geq 0\}$.'

Proposition 2.29 $m(\{S_k\}_{k=0}^n) = \prod_{k=1}^n (\det D_{G_k})^{2(n-k+1)}$, where $\{G_k\}_{k=1}^n$ are the choice parameters of $M(\{S_k\}_{k=0}^n)$. Then, the maximum is attained only for the matrix M^o associated with the parameters$\{G_1, \cdots, G_n, 0, \cdots, 0\}$. This block matrix M_0 has the distinguishing property that it is invertible and its inverse if banded with band width n (i.e. if $(M^0)^{-1} = (M_{ij})$ then $M_{ij} = 0$ whenever $\mid i - j \mid > n$).

Proof Clearly $\det M_n(\{S_k\}_{k=0}^n) \neq 0$, in order to have a non–trivial problem. By (2.22)

$$\det\ M(\{S_k\}_{k=0}^N) = \prod_{k=1}^N (\ \det\ D_{G_k})^{2(N-k+1)}$$

and as G_k are contractions, $\det D_{G_k} \leq 1$. So, the maximum $m(\{S_k\}_{k=0}^n)$ is attained only for the matrix M^o associated by Theorem 2.14 to the parameters $\{G_1, \cdots, G_n, 0, \cdots 0\}$. For proving the last part, we take into account the form (2.20) of the Cholesky factorisation. Also using Proposition A.3, we have to analyse the matrix

$$\begin{bmatrix} I & -L(\{G_k\}_{k=1}^N)D(\{G_k\}_{k=1}^N)^{-1} \\ 0 & E_{N-1}^{-1}D(\{G_k\}_{k=1}^N)^{-1} \end{bmatrix}.$$

This form shows that it is sufficient to consider the case $N = n + 1$, and as $G_{n+1} = 0$ we have

$$L(\{G_k\}_{k=1}^{n+1})D(\{G_k\}_{k=1}^{n-1})^{-1} = (*, \cdots, *, 0) \begin{bmatrix} * & * & \cdots * & 0 \\ 0 & * & \cdots * & 0 \\ \cdot & & & \cdot \\ \cdot & & & \cdot \\ \cdot & & & \cdot \\ 0 & & \cdots & * \end{bmatrix}$$

$$= \begin{bmatrix} * & * & \cdots * & 0 \\ 0 & * & \cdots & * \\ \cdot & & & \\ \cdot & & & \\ \cdot & & & \\ 0 & & \cdots & * \end{bmatrix}$$

where the entries marked with '$*$' are of no account, but the 0 in the $(0, n+1)$th entry is exactly what we need. ■

From now on we restrict ourselves to the scalar case when μ is simply a positive measure; the associated parameters given by Theorem 2.14 are denoted by $\{g_n\}_{n\geq 1}$ and the Fourier coefficients of μ are $\{s_n\}_{n\in\mathbf{Z}}$. The first result is a characterisation of the discrete measures in terms of the parameters $\{g_n\}_{n\geq 1}$. The way indicated here for obtaining this result emphasises the role played by the Naimark dilation. We denote by δ_t the measure supported by $\{e^{it}\}$, where t is fixed in $[0, 2\pi)$.

Proposition 2.30 If μ has finite support, then there exists $N \in \mathbf{N}$ such that $| g_N |= 1$.

Proof Take $\mu = \sum\limits_{k=1}^{n} a_k \delta_{t_k}$ with $a_k > 0$. Then,

$$M_n = \sum_{k=1}^{n} \begin{bmatrix} a_k & a_k e^{it_k} & \cdots & a_k e^{int_k} \\ a_k e^{-it_k} & a_k & \cdots & a_k e^{i(n-1)t_k} \\ \cdot & \cdot & \cdots & \cdot \\ \cdot & \cdot & \cdots & \cdot \\ a_k e^{int_k} & \cdot & \cdots & a_k \end{bmatrix}$$

and the matrices in the right–hand side have rank one. Consequently, $\mathrm{rank} M_n \leq n$ and $\det M_n = 0$. In view of formula (2.22), there exists $N \leq n$ such that $| g_N |= 1$. ∎

For proving the other implication, we take a positive measure μ with $| g_N |= 1$. In this case, the space $\mathcal{K}_N$ given by (2.36) is simply $\mathbf{C}^N$ and W_N given by (2.37) is a unitary operator in $\mathcal{L}(\mathbf{C}^N)$ because $V(\{g_k\}_{k=1}^{N})$ is unitary when $| g_N |= 1$.

Lemma 2.31 The operator W_N has a simple spectrum.

Proof We will show that W_N^* has a simple spectrum by finding a cyclic vector for it. As $d_k = (1- | g_k |^2)^{\frac{1}{2}} \neq 0$ for $1 \leq k < N$, we define

$$e_1 = (1, -\frac{\bar{g}_1}{d_1}, \cdots, -\frac{\bar{g}_{N-1}}{d_1 \cdots d_{N-1}}).$$

Taking (2.38) into account, we obtain by direct computation that

$$W_N^* e_1 = (0, \cdots, 0, \frac{\bar{g}_N}{d_1 \cdots d_{N-1}})$$
$$W_N^{*k} e_1 = (0, \cdots, 0, \frac{\bar{g}_N}{d_1 \cdots d_{N-k}} *, *, \cdots, *) \text{ for } 2 \leq k \leq N-1.$$

It follows that

$$\bigvee_{k-1}^{N-1} W_N^* e_1 = \mathbf{C}^N,$$

which shows that e_1 is a cyclic vector for W_N^*.

Theorem 2.32 Let μ be a positive measure on $\mathbf{T}$ associated to the parameters $\{g_n\}_{n=1}^{\infty}$. The μ has finite support if and only if there exists $N \in \mathbf{N}$ such that $| g_N |= 1$. Moreover, $| g_N |= 1$ if and only if $\mu = \sum_{k=1}^{N} a_k \delta_{t_k}$, where

$$e^{it_k} = \lambda_k, \quad k = 1, \cdots, N$$

and $\lambda_1, \cdots, \lambda_N$ are the eigenvalues of the unitary operator W_N. Denoting by p_N the characteristic polynomial of W_N, p_N' its derivative and by $\{s_n\}_{n\in\mathbf{Z}}$ the Fourier coefficients of μ,

$$a_k = \sum_{j=1}^{N} \frac{(-1)^{N+j} s_{j-1}}{p_N'(\lambda_k)} \phi_{j,k}$$

where

$$\phi_{j,k} = \sum \lambda_{r_1} \cdots \lambda_{r_{N-j}}.$$

In the last summation, $r_1, \cdots, r_{N-j}$ are different numbers running over the set $\{1, 2, \cdots, k-1, k+1, \cdots, N\}$.

Proof When μ has finite support, by Proposition 2.30 there exist $N \in \mathbf{N}$ such that $| g_N |= 1$. If $| g_N |= 1$, let E be the spectral measure of the unitary operator W_N. Then,

$$\text{supp } E = \sigma(W_N) = \{\lambda_1, \cdots, \lambda_N\}, \lambda_k \neq \lambda_j \text{ for } k \neq j$$

(supp E denotes the support of E).

If $e_o = (1, 0, \cdots, 0) \in \mathbf{C}^N$, then, the formula $\mu_o(f) = (E(f)e_o, e_o), f \in C(\mathbf{T})$ defines a positive measure on $\mathbf{T}$ with supp $\mu_o \subseteq$ supp E, and by Theorem 2.23 this measure coincides with μ and thus it follows that μ has finite support. Consequently, when $| g_N |= 1$ we have $\mu = \sum_{k=1}^{p} a_k \delta_{tk}$ with $p \leq N, e^{itk} = \lambda_k$ and $a_k > 0$. If $p < N$, according to Proposition 2.30 there exists $p_o \leq p < N$ with $| g_{p_o} |= 1$, a contradiction which ensures that $p = N$ and

$$\text{supp } \mu = \text{ supp} \mu_o = \sigma(W_N).$$

The formulas for a_k, $k = 1, \cdots, N$ can be obtained by applying the well–known method for solving systems of equations. ■

Using Theorem 2.32 and formula (2.22) we can give several characterisations for the case when the trigonometric moment problem has a unique solution.

Proposition 2.33 Let $\{s_k\}_{k=0}^n$ be such that the Toeplitz matrix $M_n = M(\{s_k\}_{k=0}^n)$ is positive definite. Then the following assertions are equivalent.

(1) The trigonometric moment problem with data $\{s_k\}_{k=0}^n$ has a unique solution.

(2) There exists $1 \leq r \leq n$ such that $| g_r |= 1$, where $\{s_k\}_{k=0}^n$ are the parameters of M_n given by Theorem 2.14.

(3) rank $M_n = r$.

(4) det $M_n = 0$.

(5) det $M_r = 0$ for some $r \leq n$ but det $M_p \neq 0$ for $0 \leq p < r$.

(6) There exists $1 \leq r \leq n$ and $a_k > 0$, $| \lambda_k |= 1$, $k = 1, \cdots, r$ with $\lambda_i \neq \lambda_j$, $i \neq j$, such that

$$s_m = \sum_{k=1}^{r} a_k \lambda_k^m, \quad m = 1, \cdots, n.$$

In Theorem 2.32 the characteristic polynomials of the operators W_n given by (2.37) naturally came to our notice. We end this section with some of their elementary properties.

Fix a positive measure μ on $\mathbf{T}$ and let $\{g_n\}_{n=1}^\infty$ be its choice parameters. Consider $W_n \in \mathcal{L}(\mathbf{C}^n)$, the operators given by 2.37, which, in general, are only contractions. Their characteristic polynomials are:

$$p_n(z) = \det(zI - W_n). \tag{2.49}$$

We keep the notation $d_n = (1- | g_n |^2)^{\frac{1}{2}}$ and we define the polynomials $q_o(z) = -1$,

$$q_1(z) = \det \begin{bmatrix} z - g_1 & -d_1 \\ -d_1 & \bar{g}_1 \end{bmatrix} \qquad (2.50)_1$$

and for $n \geq 2$,

$$qn^{(z)} = \det \begin{bmatrix} z-g_1 & -d_1 g_2 & \cdots & -d_1 d_2 \cdots d_{n-1} g_n & -d_1 \cdots d_n \\ -d_1 & z+\bar{g}_1 g_2 & \cdots & & \\ \cdot & & & & \\ \cdot & & & & \\ \cdot & & & & \\ 0 & \cdots & -d_{n-1} & z+\bar{g}_{n-1} g_n & \bar{g}_{n-1} d_n \\ 0 & \cdots & 0 & -d_n & \bar{g}_n \end{bmatrix}. \qquad (2.50)_n$$

As a consequence of the structure of W_n, the following relations hold:

$$\begin{aligned} p_1(z) &= z - g_1 \\ p_n(z) &= (z + \bar{g}_{n-1} g_n) p_{n-1}(z) + d_{n-1}^2 g_n q_{n-2}(z) \\ q_n(z) &= \bar{g}_n p_n(z) + d_n^2 q_{n-1}(z). \end{aligned} \qquad (2.51)$$

But there exists a more precise connection between the polynomials p_n and q_n.

Proposition 2.34 The following relations hold:

(1) $q_n(z) = -p_n^T(z)$.

(2.) $p_{n+1}(z) = zp_n(z) - g_{n+1} p_n(z)$

$p_{n+1}^T(z) = p_n^T(z) - z\bar{g}_{n+1} p_n(z)$.

Proof First, we prove that

$$p_{n+1}(z) = zp_n(z) + g_{n+1} q_n(z) \qquad (2.52)$$

and

$$p_{n+1}^T(z) = p_n^T(z) + z\bar{g}_{n+1} q_n^T(z). \qquad (2.53)$$

For (2.52), we use (2.51) to find

$$\begin{aligned} p_{n+1}(z) - g_{n+1} q_n(z) &= (z + \bar{g}_n g_{n+1}) p_n(z) + d_n^2 g_{n+1} q_{n-1}(z) \\ &\quad - z g_{n+1} \bar{g}_n p_n(z) - g_{n+1} d_n^2 q_{n-1}(z) = zp_n(z). \end{aligned}$$

The relation (2.53) follows from (2.52). Then, we prove (1) by induction. For $n = 1$, the relation is obviously true. By (2.51) and (2.53),

$$q_n(z) = \bar{g}_n p_n(z) + d_n^2 q_{n-1}(z)$$

$$= z\bar{g}_n p_{n-1}(z) + \mid g_n \mid^2 q_{n-1}(z) + q_{n-1}(z) - \mid g_n \mid^2 q_{n-1}(z)$$
$$= -z\bar{g}_n q_{n-1}^T(z) - p_{n-1}^T(z)$$
$$= -p_n^T(z).$$

Part (2) follows by (1), (2.52) and (2.53).

Notes

The completion idea is used in [1], and is generally developed in [9] for the structure of contractive intertwining dilations. Theorem 2.1 is a variant of the generic case of the result in [9] and follows [24]. In Chapter 6 we will present the general case of [9]. The same paper [24] contains Lemma 2.9 and Proposition 2.10. A detailed study of $\mathcal{S}(\mathcal{H}_1, \mathcal{H}_2)$, where $\mathcal{H}_1, \mathcal{H}_2$ are separable Hilbert spaces, is undertaken in [84]; Proposition 2.8 appears there.

The results concerning completion problems, such as Proposition 2.3 or the Carathéodory–Fejér Theorem 2.4 (see [19]) are described in [61]. Theorem 2.7 is taken from [43] and Proposition 2.29 is a particular case of a result in [40]. The structure of positive Toeplitz block matrices as established in Theorem 2.14 is in [22].

The determinantal formula (2.28) also appears as a matricial version of some results in orthogonal polynomials on the unit circle; see [48] and [13]. The structure of the Naimark dilation as it appears in Theorem 2.23 was established in [23]. Its isometric part appeared earlier in [21] and [55]. Results such as Theorem 2.32 and Proposition 2.33 are taken from [61]. The polynomials given by 2.49 are considered in [22] and [55].

CHAPTER 3

Spectral factorisations

Factorisations play a central role in our considerations. We have already met the Cholesky factorisation of a positive Toeplitz matrix and the main result of this chapter will be the analogue for positive definite Toeplitz kernels on the set of integers. Some computations based on the choice parameters will be presented, together with some approximation properties. Finally, a few classes of factorable functions will be analysed.

3.1 Factorisation of semispectral measures

The objective of this section is to introduce the spectral factor of a semispectral measure on the unit circle. This is developed using the geometric method of Lowdenslager.

We begin with the classical results concerning this topic. First of all we have the factorisation of positive trigonometric polynomials.

Theorem 3.1 Any trigonometric polynomial

$$P(e^{it}) = \sum_{k=0}^{n} (a_k e^{-ikt} + \bar{a}_k e^{ikt}), \quad a_n \neq 0$$

which is positive (i.e. $P(e^{it}) \geq 0$ for $t \in [0, 2\pi)$) admits the factorisation $p(e^{it}) = | Q(e^{it}) |^2$, where $Q(z) = \sum\limits_{k=0}^{n} b_k z^k$ is an analytic trigonometric polynomial which has no zeros on the unit disc.

Proof $e^{int} P(e^{it})$ extends to an algebraic polynomial of degree $2n$, so that we can write:

$$P(e^{it}) = a_n e^{-int} \prod_{k=1}^{p} (e^{it} - z_k)(e^{it} - \frac{1}{\bar{z}_k}) \prod_{j=1}^{m} (e^{it} - \xi_j)$$

where $| z_k |> 1$ for $k = 1, \cdots, p, | \xi_j |= 1$ for $j = 1, \cdots, m$ and $2p + m = 2n$.

If $s = n - p = m/2$, then

$$P(e^{it}) = \prod_{k=1}^{p}(e^{-it} - z_k)(e^{-it} - \bar{z}_k)(ae^{-its}\prod_{j=1}^{2s}(e^{it} - \xi_j))$$
$$= A \mid \prod_{k=1}^{p}(e^{it} - z_k) \mid^2 \mid \prod_{j=1}^{s}(e^{it} - \xi_j) \mid^2$$

where a and A are obvious positive constants. Thus the polynomial

$$Q(z) = A^{\frac{1}{2}} \prod_{k=1}^{p}(z - z_k) \prod_{j=1}^{s}(s - \xi_j)$$

has the required properties. ∎

This result has an extension to Hardy spaces.

Theorem 3.2 Let $f \in L^1$, $f \geq 0$. Then, there exists an outer function $G \in H^2$ with $f(t) =\mid G(e^{it}) \mid^2$ a.e. on c**T** with respect to the Lebesque measure if and only if $\log f \in L^1$.

Proof The condition is necessary: G belongs to H^2 so $\log\mid G \mid \in L^1$.

Conversely, let us define

$$G(z) = \exp\left(\frac{1}{4\pi}\int_0^{2\pi} \frac{e^{it} + z}{e^{it} - z} \log f(t)dt\right).$$

The integral is well defined by Jensen's inequality and G is actually analytic in **D** and belongs to H^2. It is known that a function of the type of G is outer and by Fatou's theorem,

$$f(t) =\mid G(e^{it}) \mid^2 \text{ a.e.on } \mathbf{T}.$$ ∎

As a consequence of this theorem we have the following result which is the scalar version of the main result in this section.

Corollary 3.3 Let μ be a positive measure on **T** with $\log d\mu/dt \in L^1$. Then there exists an outer function $G \in H^2$ such that

$$\mid G \mid^2 . \text{(Lebesgue measure)} \leq \mu$$

and G is maximal with this property in the sense that if F is another outer function satisfying the above inequality, then $| F(e^{it}) | \leq | G(e^{it}) |$ a.e. on $\mathbf{T}$.

Proof Let $f = d\mu/dt$, then $f \in L^1$, $f \geq 0$ and $\log f \in L^1$; in view of Theorem 3.2,

$$G(z) = \exp \left(\frac{1}{4\pi} \int_0^{2\pi} \frac{e^{it} + z}{e^{it} - z} \log f(t) dt\right) \tag{3.1}$$

is an outer function in H^2 satisfying $| G(e^{it}) |^2 = f(t)$; consequently

$$| G |^2 \text{ . (Lebesgue measure) } \leq \mu.$$

Now let F be another outer function with the property $| F |^2$ (Lebesgue measure) $\leq \mu$. As the continuous function $\mathrm{Re}(e^{it} + z)/(e^{it} - z)$ is also positive for fixed $z \in \mathbf{D}$, it follows that

$$\frac{1}{2\pi} \int_0^{2\pi} \mathrm{Re} \frac{e^{it} + z}{e^{it} - z} | F |^2 (e^{it}) dt \leq \frac{1}{2\pi} \int_0^{2\pi} \mathrm{Re} \frac{e^{it} + z}{e^{it} - z} d\mu(t)$$

and by Fatou's theorem,

$$| F |^2 (e^{it}) \leq d\mu/dt(t) = | G |^2 (e^{it}) \quad \text{a.e.on } \mathbf{T}.$$

■

Remark 3.4 Of course, the above G in Corollary 3.3 is uniquely determined by its maximality property, up to a multiplicative constant. Indeed, if G and G' are two maximal outer functions as in the statement of Corollary 3.3, then $| G(e^{it}) | = | G'(e^{it}) |$ a.e. on $\mathbf{T}$ and $G(z) = cG'(z), | c | = 1$ for $z \in \mathbf{D}$ by the usual properties of outer functions. The function G given by (3.1) will be called the spectral factor of μ and will usually be denoted by G_μ.

Remark 3.5 Let μ be a positive measure on $\mathbf{T}$ with $\log d\mu/dt \in L^1$ (and also suppose that $\mu(1) = 1$). By (1.20) and the remark before Theorem 1.6, we associated with μ a function F in the Schur class. We immediately get by Fatou's theorem, that

$$\begin{aligned} d\mu/dt &= \mathrm{Re} \frac{1 + e^{it} F(e^{it})}{1 - e^{it} F(e^{it})} \\ &= \frac{1 - | F(e^{it}) |^2}{| 1 - e^{it} F(e^{it}) |^2} \quad \text{a.e.on } \mathbf{T}. \end{aligned}$$

As $F \in \mathcal{S}$, the function $1 - | F(e^{it}) |^2$ is positive; as $1 - zF(z)$ is outer it follows that $\log(1 - | F(e^{it}) |^2) \in L^1$. Finally, $1 - | F(e^{it}) |^2$ is bounded (therefore in L^1) and by

Theorem 3.2, there exists an outer function $G_F \in H^1$ (and in view of the boundedness of $(1- \mid F(e^{it}) \mid^2)$, $G_F \in H^\infty$) such that $1- \mid F(e^{it}) \mid^2 = \mid G_F(e^{it}) \mid^2$. Consequently, $G\mu$ can also be chosen by the formula $G_\mu(z) = G_F(z)(1 - zF(z))^{-1}$. ∎

For semispectral measures on $\mathbf{T}$, it is not possible to use the methods in Theorem 3.2 or Corollary 3.3. Instead we have to develop some geometrical methods based on the Wold decomposition of an isometry. In order to connect the spectral factorisation of a semispectral measure with the Cholesky factorisation, we present here some models for families of isometries resembling the Wold decomposition of a single isometry.

We fix $N \in \mathbf{Z} \cup \{\infty\}, M \in \mathbf{Z} \cup \{-\infty\}, M \leq N$ and we define $\mathbf{I} = \{n \in \mathbf{Z} \mid M - 1 < n < N + 1\}, \mathbf{J} = \{n \in \mathbf{Z}/M - 1 < n < N\}$. Given a family of Hilbert spaces $\{\mathcal{E}_n\}_{n \in \mathbf{I}}$ and a family of isometries $\{v_n\}_{n \in \mathbf{I}}$, $v_n \in \mathcal{L}(\mathcal{E}_{n+1}, \mathcal{E}_n)$, we define for $n \in \mathbf{J}$,

$$\mathcal{L}_n = \mathcal{E}_n \ominus v_n \mathcal{E}_{n+1} \tag{3.2}$$

and we discern two cases as follows.

Case 1: $N \in \mathbf{Z}$, then we also define $\mathcal{L}_N = \mathcal{E}_N$ and for any $n \in \mathbf{I}$,

$$\mathcal{E}_n = \mathcal{L}_n \bigoplus v_n \mathcal{L}_{n+1} \quad \bigoplus \cdots \bigoplus v_n \cdots v_{N-1} \mathcal{L}_N. \tag{3.3}$$

This equality leads to the following model for the family $\{v_n\}_{n \in \mathbf{J}}$: define for $n \in \mathbf{J}$ the isometries

$$S_{+,n} : \bigoplus_{k=n+1}^{N} \mathcal{L}_k \to \bigoplus_{k=n}^{N} \mathcal{L}_k$$
$$S_{+,n}(e_{n+1}, e_{n+2}, \cdots, e_N) = (0, e_{n+1}, \cdots, e_N) \tag{3.4}$$

and for $n \in \mathbf{I}$, the unitary operators

$$\phi_n : \mathcal{E}_n \to \bigoplus_{k=n}^{N} \mathcal{L}_k$$
$$\phi_n(e_n, v_n e_{n+1}, \cdots, v_n \cdots v_{N-1} e_N) = (e_n, e_{n+1}, \cdots, e_N). \tag{3.5}$$

Then,

$$\phi_n v_n = S_{+,n} \phi_{n+1}. \tag{3.6}$$

Case 2: $N = \infty$, then for any $n > M - 1$, $p \geq 0$,

$$\mathcal{E}_n = (\mathcal{L}_n \bigoplus v_n \mathcal{L}_{n+1} \bigoplus \cdots \bigoplus v_n \cdots v_{n+p-1} \mathcal{L}_{n+p}) \bigoplus v_n \cdots v_{n+p} \mathcal{E}_{n+p+1}. \tag{3.7}$$

Let, for $n > M - 1$

$$\mathcal{R}_n = \bigcap_{p=0}^{\infty} v_n \cdots v_{n+p}\mathcal{E}_{n+p+1} \tag{3.8}$$

be the residual spaces. In view of (3.7), we have

$$\mathcal{E}_n = \bigoplus_{k=0}^{\infty} v_n \cdots v_{n+k-1}\mathcal{L}_{n+k} \bigoplus \mathcal{R}_n. \tag{3.9}$$

(Of course, for $k = 0, v_n \cdots v_{n+k-1}\mathcal{L}_{n+k}$ is interpreted as $\mathcal{L}_n$.) Now we are led to the following model for the family $\{v_n\}_{n\in\mathbf{J}}$: define for $n > M - 1$ the isometries

$$\begin{aligned} &S_{+,n} : \bigoplus_{k=n+1}^{\infty} \mathcal{L}_k \to \bigoplus_{k=n}^{\infty} \mathcal{L}_k \\ &S_{+,n}(e_{n+1}, e_{n+2}, \cdots) = (0, e_{n+1}, e_{n+2}, \cdots) \end{aligned} \tag{3.10}$$

and the unitary operators

$$\begin{aligned} &\phi_n : \bigoplus_{k=0}^{\infty} v_n \cdots v_{n+k-1}\mathcal{L}_{n+k} \to \bigoplus_{k=0}^{\infty} \mathcal{L}_{n+k} \\ &\phi_n(e_n, v_n e_{n+1}, \cdots) = (e_n, e_{n+1}, \cdots). \end{aligned} \tag{3.11}$$

With respect to the decomposition (3.9), v_n can be written as

$$v_n = v_n^{(1)} \bigoplus v_n^{(2)} \tag{3.12}$$

such that $v_n^{(2)} : \mathcal{R}_{n+1} \to \mathcal{R}_n$ are unitary operators and

$$\tilde{\phi}_n v_n^{(1)} = S_{+,n}\phi_{n+1}. \tag{3.13}$$

Indeed, the subspaces $\{v_n \cdots v_{n+p}\mathcal{E}_{n+p+1}\}_{p\geq 0}$ form a decreasing family, so that $v_n\mathcal{R}_{n+1} = \mathcal{R}_n$ and $\mathcal{R}_{n+1}$ reduces v_n to a unitary operator $v_n^{(2)}$. Now, (3.13) simply follows in the same way as (3.6).

We can now prove the main result of this section, i.e. an analogue of Corollary 3.3 for semispectral measures on $\mathbf{T}$. The Hardy class $H^2(\mathcal{H})$, where $\mathcal{H}$ is a Hilbert space, has previously been defined in Section 1.3. The same defifnition makes sense when $\mathcal{H}$ is replaced by a Banach space $\mathcal{X}$, such that an analytic function F on $\mathbf{D}$ belongs to $H^2(\mathcal{X})$ if

$$\sup_{0\leq r<1} \frac{1}{2\pi} \int_0^{2\pi} \| F(re^{it}) \|^2 \, dt < \infty.$$

Of special interest is the case when $\mathcal{X} = \mathcal{L}(\mathcal{H}_1,\mathcal{H}_2), \mathcal{H}_1,\mathcal{H}_2$ being two separable Hilbert spaces. The first problem here is that Fatou's theorem fails in the strong operatorial sense, as the following example shows.

Example 3.6 Take $\mathcal{H}_1 = H^2, \mathcal{H}_2 = \mathbf{C}$ and

$$F(z)f = f(z), \quad f \in H^2, z \in \mathbf{D}.$$

For every fixed $z \in \mathbf{D}$ and $f \in H^2$, take the Taylor series of f, $f(z) = \sum_{n=0}^{\infty} c_n(f)z^n$ and

$$\| F(z)f \| =| f(z |=| \sum_{n=0}^{\infty} c_n(f)z^n |\leq (\sum_{n=0}^{\infty} | c_n(f) |^2)^{\frac{1}{2}}(\sum_{n=0}^{\infty} | z^n |^2)^{\frac{1}{2}}$$
$$= (1- | z |^2)^{-\frac{1}{2}} \| f \|_{H^2}$$

so that $\| F(z) \|\leq (1- | z |^2)^{-\frac{1}{2}}, z \in \mathbf{D}$. Defining $F_n f = c_n(f), n \geq 0$, where F_n are constructions, $F(z) = \sum_{n=0}^{\infty} F_n z^n$ and F is analytic in $\mathbf{D}$. Then for $f \in H^2$ with $\| f \|_{H^2}= 1$,

$$\frac{1}{2\pi}\int_0^{2\pi} \| F(re^{it}) \|^2 dt \leq \frac{1}{2\pi}\int_0^{2\pi} \| F(re^{it})f \|^2 dt = \frac{1}{2\pi}\int_0^{2\pi} | f(re^{it}) |^2 dt$$
$$\leq \| f \|_{H^2}= 1.$$

Consequently, $F \in H^2(\mathcal{L}(H^2, \mathbf{C}))$.

Take σ to be a set of zero Lebesgue measure such that

$$F(re^{it})f = f(re^{it})$$

has a limit for $r \to 1$, for $t \notin \sigma$, and every $f \in H^2$, which is a contradiction. ∎

As a consequence of this example, we need to formulate the inequality in Corollary 3.3 in other terms. Begin with $G \in H^2(\mathcal{L}(\mathcal{H}_1,\mathcal{H}_2))$ and let $G(z) = \sum_{n=0}^{\infty} C_n z^n$ be its Taylor series. Define the table $(T_G)_{ij} \in \mathcal{L}(\mathcal{H}_1,\mathcal{H}_2)$ for $i,j \in \mathbf{Z}$ by $(T_G)_{ij} = G_{i-j}$ for $i \geq j$ and $(T_G)_{ij} = 0$ elsewhere. Moreover, we denote the i–th column in this table by $\text{col}_i T_G$. We can view $\text{col}_i T_G$ as an operator acting between $\mathcal{H}_1$ and $\bigoplus_{n\in\mathbf{Z}} \mathcal{H}_2$ and

$$\| (\text{col}_i T_G)h \|^2= \sum_{n=0}^{\infty} \| C_n h \|^2 .$$

But, as $G \in H^2(\mathcal{L}(\mathcal{H}_1, \mathcal{H}_2))$, $\sum_{n=0}^{\infty} \| C_n g \|^2 \leq M \| h \|^2$ and $\mathrm{col}_i T_G$ are bounded operators for every $i \in \mathbf{Z}$. Finally, we define the kernel

$$M_G(i,j) = (\mathrm{col}\ _jT_G)^* \ (\mathrm{col}\ _iT_G), \quad i,j \in \mathbf{Z} \tag{3.14}$$

which is obviously Toeplitz and positive definite. As a consequence of Remark 2.24, we consider μ_G, the semispectral measure on $\mathbf{T}$ having M_G as the form of the Fourier coefficients.

We have to adapt the notion of outerness by requiring that

$$\bigvee_{n=0}^{\infty} (\mathrm{col}\ _nT_G)\mathcal{H}_1 = \bigoplus_{n \geq 0} \mathcal{H}_2. \tag{3.15}$$

When G is bounded this definition coincides with the one in Section 1.3.

With these preliminaries we can state and prove the main result of this section.

Theorem 3.7 Let μ be a semispectral measure on $\mathbf{T}$ with values in $\mathcal{L}(\mathcal{H})$, $\mathcal{H}$ being a separable Hilbert space. Then, there exists an outer function G in $H^2(\mathcal{L}(\mathcal{H}, \mathcal{G}))$, where $\mathcal{G}$ is another separable Hilbert space, having the properties:

(i) $\mu_G \leq \mu$

(ii) as soon as a function F in $H^2(\mathcal{L}(\mathcal{H}, \mathcal{F}))$ satisfies $\mu_F \leq \mu$ then $\mu_F \leq \mu_G$.

Proof From the very beginning we can suppose $\mu(1) = I_{\mathcal{H}}$. Moreover, we have the relations $S_n = \mu(e^{-int}) = P_{\mathcal{H}}^{\mathcal{K}} W^n \mid \mathcal{H}$, $n \geq 0$, $\{S_n\}_{n \in \mathbf{Z}}$ being the Fourier coefficients of μ, and W the Naimark dilation of μ. It is convenient to take into consideration the unitary operator $\tilde{W}^* : \tilde{\mathcal{K}} \to \tilde{\mathcal{K}}$ where

$$\tilde{\mathcal{K}} = \cdots \bigoplus \mathcal{D}_*(\tilde{L}) \bigoplus \mathcal{H} \bigoplus \mathcal{D}_{G_1^*} \bigoplus \mathcal{D}_{G_2^*} \bigoplus \cdots \tag{3.16}$$

and

$$\tilde{W}^* = W(\{G_n^*\}_{n=1}^{\infty}), \tag{3.17}$$

and of course $\tilde{L} = \tilde{L}(\{G_n\}_{n=1}^{\infty})$.

Defining the space

$$\mathcal{K}_- = \bigvee_{n=0}^{\infty} \tilde{W}^{n*}\mathcal{H} = \mathcal{H} \bigoplus \mathcal{D}_{G_1^*} \bigoplus \mathcal{D}_{G_2^*} \bigoplus \cdots, \tag{3.18}$$

the operator

$$W_- : \mathcal{K}_- \to \mathcal{K}_-$$
$$W_- = \tilde{W}^* \mid \mathcal{K}_- \tag{3.19}$$

is an isometry; for the family of isometries $v_n = W_-$, $n \in \mathbf{Z}$, we write the Wold decomposition (3.12):

$$\mathcal{K}_- = \bigoplus_{n=0}^{\infty} \tilde{W}^{*n}\mathcal{G}_- \bigoplus \mathcal{R}_- \tag{3.20}$$

where

$$\mathcal{G}_- = \mathcal{K}_- \ominus W_-\mathcal{K}_- \tag{3.21}$$

and

$$\mathcal{R}_- = \bigcap_{n=0}^{\infty} \tilde{W}^{*n}\mathcal{K}_-. \tag{3.22}$$

In view of the minimality of $\tilde{W}$, we obtain

$$\mathcal{K} = \bigoplus_{n=-\infty}^{\infty} \tilde{W}^{*n}\mathcal{G}_- \bigoplus \mathcal{R}_- \tag{3.23}$$

and it is obvious that $\bigoplus_{n=-\infty}^{\infty} \tilde{W}^{*n}\mathcal{G}_-$ reduces $\tilde{W}$. In view of the form of $\tilde{W}^*$, we also have

$$\mathcal{G}_- = \tilde{W}^*(\cdots 0 \bigoplus \mathcal{D}_*(\tilde{L}) \bigoplus 0_{\mathcal{H}} \bigoplus 0 \bigoplus \cdots) \tag{3.24}$$

and we define

$$\mathcal{G} = \cdots 0 \bigoplus \mathcal{D}_*(\tilde{L}) \bigoplus 0_{\mathcal{H}} \bigoplus 0 \bigoplus \cdots \tag{3.25}$$

and

$$G(z) = P_{\mathcal{G}}^{\tilde{\mathcal{K}}} \tilde{W}(I - z\tilde{W})^{-1} \mid \mathcal{H}, \quad z \in \mathbf{D} \tag{3.26}$$

In the remaining part of the proof we show that this function has the required properties. We remark that G is analytic in $\mathbf{D}$ and has the Taylor series

$$G(z) = \sum_{n=0}^{\infty} z^n P_{\mathcal{G}}^{\tilde{\mathcal{K}}} \tilde{W}^{n+1} \mid \mathcal{H}. \tag{3.27}$$

This also shows that G is outer. The main point is the relation between G and the projection onto $\bigoplus_{n=-\infty}^{\infty} \tilde{W}^{*n}\mathcal{G}_-$. Denoting this projection by P, we get

$$P = \sum_{n=-\infty}^{\infty} \tilde{W}^{*n+1} P_{\mathcal{G}}^{\tilde{\mathcal{K}}} \tilde{W}^{n+1} \tag{3.28}$$

and, remarking that $P_{\mathcal{G}}^{\tilde{\mathcal{K}}} \tilde{W}^{*n} \mid \mathcal{H} = 0$ for $n > 0$, we have

$$P \mid \mathcal{H} = \sum_{n=0}^{\infty} \tilde{W}^{*n+1} P_{\mathcal{G}}^{\tilde{\mathcal{K}}} \tilde{W}^{n+1} \mid \mathcal{H}. \tag{3.29}$$

Consequently,

$$(\text{col}_i T_G)^* \ (\text{col}_j T_G) = P_{\mathcal{H}}^{\tilde{\mathcal{K}}} \tilde{W}^{*j-i} P \mid \mathcal{H} = P_{\mathcal{H}}^{\tilde{\mathcal{K}}} P \tilde{W}^{*j-i} \mid \mathcal{H}. \tag{3.30}$$

Now, we first deduce that for $i \in \mathbf{Z}$,

$$\sum_{n=0}^{\infty} \| P_{\mathcal{G}}^{\tilde{\mathcal{K}}} \tilde{W}^{n+1} h \|^2 = \| \ (\text{col}\ _i T_G) h \|^2 = \| Ph \|^2 \leq \| h \|^2$$

ensuring that $G \in H^2(\mathcal{L}(\mathcal{H}, \mathcal{G}))$ and then, for any $n \in \mathbf{N}$ and $h_o, \cdots .h_n$ in $\mathcal{H}$,

$$\begin{aligned}
\sum_{i,j=0}^{n} (M_G(i,j) h_j, h_i) &= \sum_{i,j=0}^{n} \ ((\text{col}\ _j T_G)^* \ (\text{col}\ _i T_G) h_j, h_i) \\
&= \sum_{i,j=0}^{n} (\tilde{W}^{j-1} P h_j, h_i) = \| P \sum_{k=0}^{n} \tilde{W}^k h_k \|^2 \\
&\leq \| \sum_{k=0}^{n} \tilde{W}^k h_k \|^2 = \sum_{i,j=0}^{n} (S_{j-i} h_j, h_i)
\end{aligned}$$

proving the inequality (i).

Let F in $H^2(\mathcal{L}(\mathcal{H}, \mathcal{F}))$ be another function satisfying $\mu_F \leq \mu$. We define:

$$X : \mathcal{K}_- \to \bigoplus_{n=0}^{\infty} \mathcal{F}$$

$$X \sum_{k=0}^{n} \tilde{W}^{*k} h_k = \sum_{k=0}^{n} \ (\text{col}\ _k T_F) h_k$$

and we get

$$\begin{aligned}
\| X\sum_{k=0}^{n} \tilde{W}^{*k}h_k \|^2 &= \| \sum_{k=0}^{n} (\mathrm{col}\ _kT_F)h_k \|^2 = \sum_{i,j=0}^{n} ((\mathrm{col}\ _iT_F)^* (\mathrm{col}\ _jT_F)h_j, h_i) \\
&= \sum_{i,j=0}^{n} (M_F^*(i,j)h_j, h_i) = \sum_{i,j=0}^{n} (h_j, M_F(i,j)h_i) \\
&\leq \sum_{i,j=0}^{n} (h_j, S_{j-i}h_i) = \sum_{i,j=0}^{n} (h_j, \tilde{W}^{j-i}h_i) = \sum_{i,j=0}^{n} (\tilde{W}^{*j-1}h_j, h_i) \\
&= \| \sum_{k=0}^{n} \tilde{W}^{*k}h_k \|^2
\end{aligned}$$

and X extends to a contraction between $\mathcal{K}_-$ and $\bigoplus_{n=0}^{\infty} \mathcal{F}$.

On the one hand, we observe that

$$\begin{aligned}
X\tilde{W}^*\left(\sum_{k=0}^{n} \tilde{W}^{*k}h_k\right) &= X\left(\sum_{k=1}^{n+1} \tilde{W}^{*k}h_{k-1}\right) = \sum_{k=1}^{n+1} (\mathrm{col}_kT_F)h_{k-1} \\
&= S_{+,0}\left(\sum_{k=0}^{n} (\mathrm{col}_kT_F)h_k\right) = S_{+,0}X\left(\sum_{k=0}^{n} \tilde{W}^{*k}h_k\right)
\end{aligned}$$

where $S_{+,0}$ is the corresponding translation operator given by (3.10); consequently,

$$X\tilde{W}^* = S_{+,0}X. \tag{3.32}$$

On the other hand, it follows that:

$$X\mathcal{R}_- = X\bigcap_{n=0}^{\infty} \tilde{W}^{*n}\mathcal{K}_- \subset \bigcap_{n=0}^{\infty} X\tilde{W}^{*n}\mathcal{K}_- = \bigcap_{n=0}^{\infty} S_{+,0}^n X\mathcal{K}_- \subset \bigcap_{n=0}^{\infty} S_{+,0}^n \left(\bigoplus_{m=0}^{\infty} \mathcal{F}\right) = 0$$

by an obvious property of the translation operator. The relation $X\mathcal{R}_- = 0$ can be rewritten as $XP = X$ and finally, we have for $n \in \mathbf{N}$ and $h_o, \cdots, h_n$ in $\mathcal{H}$:

$$\begin{aligned}
\sum_{i,j=0}^{n} (M_F^*(i,j)h_j, h_i) &= \| \sum_{k=0}^{n} (\mathrm{col}_kT_F)h_k \|^2 = \| X\sum_{k=0}^{n} \tilde{W}^{*k}h_k \|^2 \\
&= \| XP\sum_{k=0}^{n} \tilde{W}^{*k}h_k \|^2 \leq \| P\sum_{k=0}^{n} \tilde{W}^{*k}h_k \|^2 = \sum_{i,j=0}^{n} (M_G^*(i,j)h_j, h_i).
\end{aligned}$$

Consequently, (ii) is also true. ∎

Two questions remain to be elucidated; the first one concerns the conditions necessary to have equality in the statement (i) of Theorem 3.7 and the second one concerns the uniqueness of the function G satisfying (i) and (ii).

Proposition 3.8 The equality in (i) of Therem 3.7 holds if and only if $\mathcal{R}_- = \bigcap_{n=0}^{\infty} \tilde{W}^{*n}\, \mathcal{K}_- = 0$.

Proof Analysing the proof of (i) from Theorem 3.7, we get the result in an obvious way. ∎

In order to obtain a uniqueness property of G as is the case in Remark 3.4, we need a factorisation result which is a variant 'with functions' of Proposition A.2.

Proposition 3.9 Let $F \in H^2(\mathcal{L}(\mathcal{H},\mathcal{F}))$ and $F' \in H^2(\mathcal{L}(\mathcal{H},\mathcal{F}'))$ be two functions such that $M_F \leq M_{F'}$ and the second one is assumed to be outer. Then there exists a function $F'' \in \mathcal{S}(\mathcal{F}',\mathcal{F})$ such that $F(z) = F''(z)F'(z)$ for $z \in \mathbf{D}$. When $M_F = M_{F'}$, F'' is inner and if in addition F is supposed to be outer, then F'' is a unitary constant function.

Proof We define the operator

$$Q : \bigoplus_{n=0}^{\infty} \mathcal{F}' \to \bigoplus_{n=0}^{\infty} \mathcal{F}$$

$$Q(\sum_{k=0}^{n} (\mathrm{col}_k T_{F'})h_k) = \sum_{k=0}^{n} (\mathrm{col}_k T_F)h_k \tag{3.33}$$

and $M_F \leq M_{F'}$ which means precisely that Q extends to a contraction between $\bigvee_{n=0}^{\infty} (\mathrm{col}_n T_{F'})\mathcal{H}$ and $\bigvee_{n=0}^{\infty} (\mathrm{col}_n T_F)\mathcal{H}$. In view of the outerness of F', $\bigvee_{n=0}^{\infty} (\mathrm{col}_n T_{F'})\mathcal{H} = \bigoplus_{n=0}^{\infty} \mathcal{F}'$. In the same way as we obtained (3.32), we deduce that

$$QS'_{+,0} = S_{+,0}Q \tag{3.34}$$

where $S'_{+,0}$ and $S_{+,0}$ are the corresponding translation operators.

Relation (3.34) implies that Q has a lower triangular Toeplitz matrix with respect to the decompositions $\bigoplus_{n=0}^{\infty} \mathcal{F}'$ and $\bigoplus_{n=0}^{\infty} \mathcal{F}$, whose symbol is an analytic bounded function

$F'' \in \mathcal{S}(\mathcal{F}', \mathcal{F})$. Moreover, for $h \in \mathcal{H}$,

$$Q\ (\mathrm{col}_o T_{F'})h = (\mathrm{col}_o T_F)h$$

which is exactly the equality $F(z)h = F''(z)F'(z)h$, $z \in \mathbf{D}$, $h \in \mathcal{H}$.

When $M_F = M_{F'}$, Q is an isometry and by definition, F'' is inner. If F is also supposed to be outer, then F'' is outer and an inner and outer function in the class $\mathcal{S}(\mathcal{F}', \mathcal{F})$ is a unitary constant. ∎

Now, we can obtain the operatorial variant of Remark 3.4 regarding the uniqueness of G.

Proposition 3.10 Let μ be a semispectral measure on $\mathbf{T}$ with values in $\mathcal{L}(\mathcal{H})$. Then, an outer function in $H^2(\mathcal{L}(\mathcal{H}, \mathcal{G}))$ satisfying the conditions (i) and (ii) in the statement of Theorem 3.7 is uniquely determined up to a multiplicative unitary constant factor on the left.

Proof Let G and G' be two outer factors satisfying (i) and (ii) in the statement of Theorem 3.7. In particular, $M_G = M_{G'}$ and by Proposition 3.9, G and G' differ by a unitary constant factor on the left.

In view of Proposition 3.10 we will call the function G produced by Theorem 3.7 the spectral factor of the semispectral measure μ and it is usually denoted by G_μ.

Remark 3.11 It remains to explain the connection between the procedure of obtaining the spectral factor of μ and the Cholesky factorisation of $M_n(\mathcal{T})$, where $\mathcal{T}$ is the Toeplitz form based on the Fourier coefficients of μ.

First of all, we note that the spectral factor produces an upper–lower triangular factorisation, while the Cholesky factorisation $(2.25)_n$ produces a lower–upper triangular factorisation, so that we also have to take into account an upper–lower triangular Cholesky factorisation. For this, we define $\hat{E}_o = I_{\mathcal{H}}$ and for $n \geq 1$,

$$\hat{E}_n : \bigoplus_{k=0}^{n} \mathcal{H} \to \mathcal{D}_{G_n^*} \bigoplus \mathcal{D}_{G_{n-1}^*} \bigoplus \cdots \bigoplus \mathcal{D}_{G_1^*} \bigoplus \mathcal{H}$$

$$\hat{E}_n = u_n E_n(\{G_k^*\}_{k=1}^n) u_n \tag{3.35}$$

where u_n are the unitary operators defined by (2.13) (of course, on corresponding spaces) and $\{G_n\}_{n=1}^{\infty}$ are the choice parameters of μ. $\hat{E}_n$ is lower triangular and

$$\hat{E}_n^* \hat{E}_n = u_n E_n^*(\{G_k^*\}_{k=1}^n) E_n(\{G_k^*\}_{k=1}^n) u_n = u_n M(\{G_k^*\}_{k=1}^n) u_n = M_n,$$

where we also used (2.25).

Now, we fix $n \in \mathbf{N}$ and we define the family of isometries:

$$\begin{aligned} v_o &= 0 : \mathcal{E}_1 = 0 \to \mathcal{E}_o = \mathcal{H} \\ v_{-1} &= (G_1, D_{G_1})^{\mathrm{t}} : \mathcal{E}_o \to \mathcal{E}_{-1} = \mathcal{K}_2 \\ &\cdot \\ &\cdot \\ &\cdot \\ v_{-n} &= (L(\{G_k\}_{k=1}^n), D(\{G_k\}_{k=1}^n))^{\mathrm{t}} : \mathcal{E}_{-n+1} \to \mathcal{E}_{-n} = \mathcal{K}_{n+1} \end{aligned}$$

where the spaces $\mathcal{K}_p$ are defined by (2.36). (The notation continues to be the same as in the Appendix and we consent to the convention that $0 : 0 \to \mathcal{H}$ is an isometry.) Then, we take into account the Wold decomposition (3.3) of this family and

$$\mathcal{L}_k = \mathcal{E}_{-k} \ominus v_k \mathcal{E}_{-k+1}, \quad k = 0, \cdots, n,$$

and in particular, $\mathcal{L}_o = \mathcal{H}$. Considering the unitary operators

$$\begin{aligned} &w_k : \mathcal{F}_{-k+1} \to \mathcal{F}_{-k} \\ &w_k = I \bigoplus \begin{bmatrix} I & 0 \\ 0 & \alpha(L\{G_m\}_{m=1}^k) \end{bmatrix} J(L\{G_m\}_{m=1}^k) \begin{bmatrix} 0 & I \\ \beta^*(L\{G_m\}_{m=1}^k) & 0 \end{bmatrix} \end{aligned} \tag{1.37}$$

where

$$\begin{aligned} \mathcal{F}_o &= \mathcal{D}_{G_n^*} \bigoplus \cdots \bigoplus \mathcal{D}_{G_1^*} \bigoplus \mathcal{H} \\ \mathcal{F}_{-1} &= \mathcal{D}_{G_n^*} \bigoplus \cdots \bigoplus \mathcal{D}_{G_2^*} \bigoplus \mathcal{H} \bigoplus \mathcal{D}_{G_1} \\ &\cdot \\ &\cdot \\ &\cdot \\ \mathcal{F}_{-n} &= \mathcal{K}_{n+1} \end{aligned} \tag{1.38}$$

we get

$$\mathcal{L}_{-k} = w_{-k}(\cdots 0 \bigoplus \mathcal{D}_{G_k^*} \bigoplus \boxed{0} \bigoplus 0 \bigoplus \cdots), \quad 1 \leq k \leq n. \tag{1.39}$$

Denote by $\hat{E}_n^{(ij)}$, $-n \leq i,j \leq 0$, the matrix elements of E_n. Then, the point is that

$$\hat{E}_n^{(ij)} = P_{\mathcal{D}_{G^*_{-j}}} w^*_{i-1} \cdots w^*_j \mid \mathcal{H} \tag{1.40}$$

where $\mathcal{D}_{G^*_{-j}}$ are naturally identified with $(0 \bigoplus \cdots \mathcal{D}_{G^*_{-j}} \bigoplus \boxed{0} \bigoplus \cdots \bigoplus 0)$.

As by (3.3),

$$\mathcal{E}_{-n} = \mathcal{L}_{-n} \bigoplus w_{-n}\mathcal{L}_{-n+1} \bigoplus \cdots \bigoplus w_{-n} \cdots w_{-1}\mathcal{H}.$$

Equation (3.40) results from a computation similar to (3.30) because here there is no residual space. This shows that the Cholesky factor (3.35) of M_n can be obtained by the same procedure as the spectral factor of a semispectral measure on **T**.

This is the first time that a non–stationary analysis would explain the facts better. We will continue this analysis in the second volume of these notes. ∎

The last result in this section concerns the inner–outer factorisation for functions in $H^2(\mathcal{L}(\mathcal{H}_1, \mathcal{H}_2))$.

Proposition 3.12 Any function F in $H^2(\mathcal{L}(\mathcal{H}_1, \mathcal{H}_2))$ admits a uniquely determined (up to multiplicative unitary factors) factorisation $F = F_i F_o$, with F_o an outer function in $H^2(\mathcal{L}(\mathcal{H}_1, \mathcal{H}_3))$ for a certain separable Hilbert space $\mathcal{H}_3$, and F_i an inner function with values in $\mathcal{L}(\mathcal{H}_3, \mathcal{H}_2)$.

Proof Let μ_F be the semispectral measure on **T** associated with F; then by (3.14) and Theorem 3.7, consider its spectral factor, denoted here by F_o, an outer function in $H^2(\mathcal{L}(\mathcal{H}_1, \mathcal{H}_3))$. Consequently, $M_F = M_{F_o}$ and by Proposition 3.9, there exists an inner function F_i such that $F = F_i F_o$.

For uniqueness, take $F = F_i' F_o'$ to be another inner–outer factorisation of F. Then, for any $n \in \mathbf{N}$ and $h_o, \cdots, h_n \in \mathcal{H}_1$,

$$\begin{aligned}\sum_{i,j=0}^{n} (M_F^*(i,j)h_j, h_i) &= \| \sum_{k=0}^{n} (\mathrm{col}_k T_F) h_k \|^2 \\ &= \| \sum_{k=0}^{n} (T_{F_i'} \mathrm{col}_k T_{F_o'}) h_k \|^2 = \| \sum_{k=0}^{n} (\mathrm{col}_k T_{F_o'}) h_k \|^2\end{aligned}$$

$$= \sum_{i,j=0}^{n} (M^*_{F'_o}(i,j)h_j, h_i)$$

and $M_F = M_{F'_o} = M_{F_o}$. Again by Proposition 3.9, $F'_o = F_i U^*$. ■

3.2 Some formulas for spectral factors

The main purpose of this section is to obtain an operatorial variant of the formula in Remark 3.5. Of course, by Theorem 3.7, for every function N on $[0, 2\pi]$ whose values are positive contractions on a separable Hilbert space $\mathcal{H}$ which is measurable (strongly or weakly, in view of the separability of $\mathcal{H}$), there exists an outer function G_N in a class $\mathcal{S}(\mathcal{H}, \mathcal{G})$ with the properties

(i′) $N^2(t) \geq G^*_N(e^{it})G_N(e^{it})$ a.e.

(ii′) for any other function G' in a class $\mathcal{S}(\mathcal{H}, \mathcal{G}')$ such that $N^2(t) \geq G'^*(e^{it})G'(e^{it})$ a.e., we have

$$G^*_N(e^{it})G_N(e^{it}) \geq G'^*(e^{it})G'(e^{it}) \text{ a.e.}$$

Indeed, as $0 \leq N^2(t) \leq I$, we simply consider the Toeplitz form determined by the Fourier coefficients of N and then the measure μ_N associated to this Toeplitz form. Applying Theorem 3.7, we get an outer function G_N. The condition $\mu_{G_N} \leq \mu_N$ will mean exactly condition (i′) above, which also ensures that G_N is actually in a Schur class, and the maximality condition (ii) in the statement of Theorem 3.7 translates into the maximality condition (ii′) for G_N.

Moreover, the properties (i′) and (ii′) determine G_N up to a constant unitary factor from the left.

In view of these considerations, for a function $F \in \mathcal{S}(\mathcal{H})$ we denote by L_F the spectral factor of $I - F(e^{it})^* F(e^{it})$. The *–outer spectral factor of the function $I - F(e^{it})F(e^{it})^*$, which is defined by an obvious dual version of (i′) and (ii′), is denoted by R_F.

Now we prove that the factorisation in Theorem 3.7 can be deduced from the factorisation of a certain function on $[0, 2\pi]$ with values of positive contractions on a separable Hilbert space and (strongly or weakly) measurable.

Proposition 3.13 Let $\mu : C(\mathbf{T}) \to \mathcal{L}(\mathcal{H})$ be a semispectral measure with $\mu(1) = I$

and

$$F(z) = \frac{1}{z}(\mu\left(\frac{e^{it}+z}{e^{it}-z}\right) - I)(\mu\left(\frac{e^{it}+z}{e^{it}-z}\right) + I)^{-1}$$

which belongs to $\mathcal{S}(\mathcal{H})$. Denote by L_F the left spectral factor or F; then the spectral factor of μ can be chosen as

$$G\mu(z) = L_F(z)(I - zF(z))^{-1}.$$

Proof Consider $L_F \in \mathcal{S}(\mathcal{H},\mathcal{G})$ the left spectral factor of F. By Proposition 3.9, $I - F(z)^*F(z) \geq L_F^*(z)L_F(z)$ for $z \in \mathbf{D}$ follows from the property (i′) of L_F and then we have:

$$\begin{aligned}
&\sup_{r<1} \frac{1}{2\pi}\int_0^{2\pi} \| L_F(-re^{it})(I - re^{it}F(re^{it}))^{-1}h \|^2\, dt \\
=&\sup_{r<1} \frac{1}{2\pi}\int_0^{1\pi} ((I - re^{-it}F^*(re^{it}))^{-1}(I - r^2F^*(re^{it})F(re^{it}))(I - re^{it}F(re^{it}))^{-1}h, h)dt \\
=&\sup_{r<1} \frac{1}{2\pi}\int_0^{2\pi} (\mu\left(\frac{1-r^2}{| e^{is} - re^{it} |^2}\right) h, h)dt \\
=&\sup_{r<1} \frac{1}{2\pi}\int_0^{2\pi}\int_0^{2\pi} \frac{1-r^2}{| e^{is} - re^{it} |^2} d\mu_h(t)dt \leq \| h \|^2
\end{aligned}$$

for every $h \in \mathcal{H}$, by Fubini's theorem and the properties of the Poisson kernel. Moreover, $\mu_h(f) = (\mu(f)h, h), f \in C(\mathbf{T}), h \in \mathcal{H}$. Consequently $L_F(I - zF)^{-1} \in H^2(\mathcal{L}(\mathcal{H},\mathcal{G}))$. By the outerness of $L_F, L_F(I - z_F)^{-1}$ is an outer function too.

Now we prove that $L_F(I - zF)^{-1}$ satisfies the properties (i) and (ii) in the statement of Theorem 3.7. First, for $h \in \mathcal{H}$ and p an analytic polynomial, we have

$$\begin{aligned}
&\frac{1}{2\pi}\int_0^{2\pi} | p(e^{it}) |^2 d(\mu_{L_F(I-zF)^{-1}}(t)h, h) \\
&\qquad \sup_{r<1} \frac{1}{2\pi}\int_0^{2\pi} | p(re^{it}) |^2 \| L_F(I - zF)^{-1}(re^{it})h \|^2\, dt \\
&\qquad \leq \sup_{r<1} \frac{1}{2\pi}\int_0^{2\pi} | p(re^{it}) |^2\, (\mu\left(\frac{1-r^2}{| e^{is} - re^{it} |^2}\right) h, h)dt \\
&\qquad = \frac{1}{2\pi}\int_0^{2\pi} | p(e^{it}) |^2\, d(\mu(t)h, h)
\end{aligned}$$

where we again used Fubini's theorem. Consequently,

$$\mu_{L_F(I-zF)^{-1}} \leq \mu.$$

Then let $H \in H^2(\mathcal{L}(\mathcal{H}, \mathcal{G}'))$ satisfy $\mu_H \le \mu$. For $h \in \mathcal{H}$ and $z \in \mathbf{D}$,

$$\begin{aligned}
\| H(z)h \|^2 &= (\mu_H \left(\frac{1 - | z |^2}{| e^{it} - z |^2} \right) h, h) \\
&\le (\mu \left(\frac{1 - | z |^2}{| e^{it} - z |^2} \right) h, h) \\
&= ((I - \bar{z}^* F(z))^{-1} (I - | z |^2 F^*(z) F(z)) (I - zF(z))^{-1} h, h).
\end{aligned}$$

Considering the function $L = H(I - zF)$, we get from the above relations that

$$L^*(z)L(z) \le I - | z |^2 F^*(z)F(z) \quad \text{for } z \in \mathbf{D},$$

and consequently, $L \in \mathcal{S}(\mathcal{H}, \mathcal{G}')$ and

$$L^*(e^{it})L(e^{it}) \le I - F^*(e^{it})F(e^{it}) \quad \text{a.e.}$$

By property (ii$'$) of L_F, we have

$$L_F^*(e^{it})L_F(e^{it}) \ge L^*(e^{it})L(e^{it}) \quad \text{a.e.},$$

hence

$$(I - \bar{z}F^*(z))^{-1} L_F^*(z) L_F(z) (I - zF(z))^{-1} \ge H^*(z)H(z) \quad \text{for } z \in \mathbf{D}$$

which will imply that $\mu_{L_F(I-zF)^{-1}} \ge \mu_H$. That is, $L_F(I - zF)^{-1}$ can be chosen as the spectral factor of μ. ■

For stating the main result of this section some more notation is necessary. Recalling the matrix form of the Naimark dilation given by Theorem 2.23, we define the operators:

$$\begin{aligned}
B &= P_{\mathcal{H}}^{\mathcal{K}} W \mid \mathcal{K} \ominus \mathcal{H} \\
C &= P_{\mathcal{K} \ominus \mathcal{H}}^{\mathcal{K}} W \mid \mathcal{H} \\
A &= P_{\mathcal{K} \ominus \mathcal{H}}^{\mathcal{K}} W \mid \mathcal{K} \ominus \mathcal{H}.
\end{aligned}$$

We have $W = \begin{bmatrix} G_1 & C \\ B & A \end{bmatrix} : \begin{matrix} \mathcal{H} \\ \oplus \\ \mathcal{K} \ominus \mathcal{H} \end{matrix} \to \begin{matrix} \mathcal{H} \\ \oplus \\ \mathcal{K} \ominus \mathcal{H} \end{matrix}$ and taking into account the dual construction presented in the proof of Theorem 3.7, we have

$$\tilde{W} = \begin{bmatrix} G_1 & \tilde{B} \\ \tilde{C} & \tilde{A} \end{bmatrix} : \begin{matrix} \mathcal{H} \\ \oplus \\ \tilde{\mathcal{K}} \ominus \mathcal{H} \end{matrix} \to \begin{matrix} \mathcal{H} \\ \oplus \\ \tilde{\mathcal{K}} \ominus \mathcal{H} \end{matrix}$$

where the rule '~' is used as in the Appendix.

The following result plays only a technical role here. As it will have an important meaning in Chapter 5, we shall postpone its proof until then. Take a semispectral measure μ with values in $\mathcal{L}(\mathcal{H})$ (our simplifying assumption $\mu(1) = I$ is also made), let $\{G_n\}_{n-1}^{\infty}$ be its parameters given by Theorem 2.14 and let F be the function in $\mathcal{S}(\mathcal{H})$ defined in the statement of Proposition 3.13.

Theorem 3.14 $F(z) = G_1 + z\tilde{B}(I - z\tilde{A})^{-1}\tilde{C}$ for $z \in$ '**D**.

Now the main result of this section is the following.

Theorem 3.15 The spectral factor of μ can be chosen as

$$G_\mu(z) = P_{\mathcal{G}}^{\tilde{\mathcal{K}}}(I - z\tilde{A})^{-1}\tilde{C}(I - zF(z))^{-1}, \quad z \in \mathbf{D},$$

where $\mathcal{G}$ is the space defined by (3.25).

Proof By (3.26), for $z \in \mathbf{D}$,

$$G_\mu(z) = P_{\mathcal{G}}^{\tilde{\mathcal{K}}}\tilde{W}(I - z\tilde{W})^{-1} \mid \mathcal{H} = P_{\mathcal{G}}^{\tilde{\mathcal{K}}}(I - z\tilde{W})^{-1}\tilde{W} \mid \mathcal{H}.$$

Taking into account the following matrix representation of $\tilde{W}$ with respect to the decomposition $\tilde{\mathcal{K}} = \mathcal{H} \oplus (\tilde{\mathcal{K}} \ominus \mathcal{H})$

$$\tilde{W} = \begin{bmatrix} G_1 & \tilde{B} \\ \tilde{C} & \tilde{A} \end{bmatrix},$$

we can use Proposition A.3 in order to get, via Theorem 3.14,

$$G\mu(z) = P_{\mathcal{G}}^{\tilde{\mathcal{K}}} \begin{bmatrix} * & * \\ z(I - z\tilde{A})^{-1}\tilde{C}\tilde{R}^{-1} & z^2(I - z\tilde{A})^{-1}\tilde{C}\tilde{R}^{-1}\tilde{B}(I - z\tilde{A})^{-1} + (I - z\tilde{A})^{-1} \end{bmatrix} \begin{bmatrix} G_1 \\ \tilde{C} \end{bmatrix}$$

where

$$\tilde{R} = I - zG_1 - z^2\tilde{B}(I - z\tilde{A})^{-1}\tilde{C} = I - zF(z).$$

The entries marked by '$*$' do not enter into computations because $\mathcal{G} \subset \tilde{\mathcal{K}} \ominus \mathcal{H}$. Furthermore,

$$G\mu(z) = P_{\mathcal{G}}^{\tilde{\mathcal{K}}}(z(I - z\tilde{A})^{-1}\tilde{C}\tilde{R}^{-1}G_1 + z^2(I - z\tilde{A})^{-1}\tilde{C}\tilde{R}^{-1}\tilde{B}(I - z\tilde{A})^{-1}\tilde{C} + (I - z\tilde{A})^{-1}\tilde{C})$$

$$= P_{\mathcal{G}}^{\tilde{\mathcal{K}}}(I - z\tilde{A})^{-1}\tilde{C}(I + z\tilde{R}^{-1}F(z)) = P_{\mathcal{G}}^{\tilde{\mathcal{K}}}(I - z\tilde{A})^{-1}\tilde{C}(I - zF(z)).$$ ∎

Corollary 3.16 Let F be a function in the Schur class $\mathcal{S}(\mathcal{H})$. Then L_F can be chosen as $L_F(z) = P_{\mathcal{G}}^{\tilde{\mathcal{K}}}(I - z\tilde{A})^{-1}\tilde{C}$, $z \in \mathbf{D}$, where the operators $\tilde{A}, \tilde{C}$ and the spaces $\tilde{\mathcal{K}}$ and $\mathcal{G}$ are associated as above with the Schur parameters of F.

Remark 3.17 We can consider $\mathcal{F} \in \mathcal{S}(\mathcal{H}_1, \mathcal{H}_2)$ for two different separable Hilbert spaces and ask if the above formuls for L_F still holds. This is indeed the case because we consider $\hat{\mathcal{H}}$ to be a Hilbert space containing both $\mathcal{H}_1$ and $\mathcal{H}_2$ and $\hat{F} \in \mathcal{L}(\hat{\mathcal{H}})$ given by the extension of F with 0. The Schur parameters of F are obviously

$$\hat{G}_n = \begin{bmatrix} G_n & 0 \\ 0 & 0 \end{bmatrix} : \begin{matrix} \bigoplus \mathcal{D}_{G_{n-1}} \\ \mathcal{H}_1 \end{matrix} \quad \to \quad \begin{matrix} \bigoplus \mathcal{D}_{G^*_{n-1}} \\ \mathcal{H}_2 \end{matrix}$$

and an inspection of the proofs above shows that the formula in Corollary 3.16 for L_F is also valid in this case. ∎

Remark 3.18 Another question concerns the determination of right spectral factor of a function $F \in \mathcal{S}(\mathcal{H}_1, \mathcal{H}_2)$. We remark by the definition that $\tilde{R}_F = L_{\tilde{F}}$ (the notation $\tilde{F}$ for $F \in \mathcal{S}(\mathcal{H}_1, \mathcal{H}_2)$ was introduced at the beginning of Section 1.3). It is obvious that if $\{G_n\}_{n=1}^{\infty}$ are the Schur parameters of F, then the Schur parameters of $\tilde{F}$ are $\{G_n^*\}_{n=1}^{\infty}$. Finally, we get (by using dual versions of the results necessary to obtain the formula in Corollary 3.16),

$$R_F(z) = P_{\mathcal{H}}^{\mathcal{K}}C(I - zA)^{-1} \mid \tilde{\mathcal{G}}, \quad z \in \mathbf{D}$$

where

$$\tilde{\mathcal{G}} = \cdots \bigoplus \mathcal{D}_*(L) \bigoplus 0_{\mathcal{H}} \bigoplus \cdots \subset \mathcal{K}$$

and of course, $L = L(\{G_n\}_{n=1}^{\infty})$.

3.3 Approximation of the spectral factors

We have seen in Section 1.3 that the operatorial Schur algorithm associates to an $F \in \mathcal{S}(\mathcal{H})$, four sequences of functions $\{a_n\}$, $\{b_n\}$, $\{c_n\}$, $\{d_n\}$ by the formula (1.38). Moreover, in the proof of Theorem 1.12, the functions a_n played a role in approximating F. More precisely, we have proved that $\{a_n\}$ converges uniformly on compact subsets of $\mathbf{D}$, in the uniform norm, to F. The purpose of this section is to show that the functions b_n and c_n play a role in the approximation of the spectral factors of F.

First, we need some preliminaries. As we will use the presence of the elements involved by the Toeplitz form associated to F (see Proposition 2.18) we consider the Schur parameters of F as being $\{G_n\}_{n=1}^{\infty}$. Then, consider the row–contraction $L = L(\{G_n\}_{n=1}^{\infty})$ and let $W = W(\{G_n\}_{n=1}^{\infty})$ be the unitary operator defined by (2.35). In the Appendix we investigate the operator

$$K_{\infty} = \alpha(L)L^*\beta^*(L) : \mathcal{D}_*(L) \to \mathcal{D}(L)$$

(where $\mathcal{D}(L)$ and $\mathcal{D}_*(L)$ are defined by Proposition A.6.)

Also in the Appendix we defined the operators

$$\begin{aligned} H_n &: \mathcal{H} \to \mathcal{D}_{G_n^*} \\ H_n &= D_{G_n^*} \cdots D_{G_1^*} \end{aligned}$$

and now define $H_n' = (H_n^* H_n)^{\frac{1}{2}}$. Moreover, consider the unitary operator

$$\begin{aligned} \beta_n' &: \mathcal{D}_{G_n^*} \to \overline{\mathcal{R}(H_n')} \\ \beta_n' H_n &= H_n'. \end{aligned} \tag{3.4}$$

Observe that $H_n' \leq H_m' \leq H_{\infty}$ for $n \geq m$, and consequently, $\overline{\mathcal{R}(H_n')} \subset \overline{\mathcal{R}(H_m')} \subset \mathcal{D}_*(L)$. One more convention is to identify $\mathcal{D}_*(L)$ with $(\cdots \mathcal{H} \bigoplus \mathcal{D}_*(L) \bigoplus 0_{\mathcal{H}} \bigoplus \cdots) = \tilde{\mathcal{G}}$, so that $\overline{\mathcal{R}(H_n')}$ can also be viewed as a subspace of $\tilde{\mathcal{G}}$. Finally, note that by (1.39) $b_n(z) = z^n v_n(z)$ with v_n a *–outer function in $\mathcal{S}(\mathcal{D}_{G_n^*}, \mathcal{H})$ (see also Proposition 2.2).

Now we can state and prove the main result of this section.

Theorem 3.19 Let F be a function in $\mathcal{S}(\mathcal{H}), L_F$ and R_F be its spectral factors and $\{G_n\}_{n=1}^{\infty}$ be its Schur parameters. Then $\{\tilde{\beta}_n' c_n\}_{n\geq 1}$ converges coefficient–wise in the strong operatorial topology to L_F, and $\{\beta_n' \tilde{v}_n\}_{n\geq 1}$ converges coefficient–wise in the strong operatorial topology to $\tilde{R}_F$.

Proof By Remark 3.18

$$\tilde{R}_F(z) = P_{\tilde{\mathcal{G}}}^{\mathcal{K}}(I - zA^*)^{-1}C^* = H_{\infty}(L) + \sum_{k=1}^{\infty} z^k P_{\tilde{\mathcal{G}}}^{\mathcal{K}} A^{*k} C^*.$$

Defining $A_+ = P_{\mathcal{K}_+ \ominus \mathcal{H}} A \mid \mathcal{K}_+ \ominus \mathcal{H}$ and $C_+ = P_{\mathcal{K}_+ \ominus \mathcal{H}} C$ we further get

$$\tilde{R}_F(z) = H_{\infty}(L) - \sum_{k=1}^{\infty} z^k K_{\infty}^* A_+^{*(k-1)} C_+^*.$$

Then, as

$$I - a_n(e^{it})Ra_n^*(e^{it}) = b_n(e^{it})b_n^*(e^{it}) \quad \text{a.e. on } \mathbf{T}$$

it follows that $R_{a_n} = v_n$, so $\tilde{R}_{a_n} = \tilde{v}_n$. On the other hand, viewing a_n as a function in $\mathcal{S}(\mathcal{H})$ associated with the Schur parameters $\{G_1, G_2, \cdots, G_n, 0, 0, \cdots\}$ (see Remark 1.13) we can again use Remark 3.18 to obtain

$$\beta_n' \tilde{v}_n(z) = H_n' - \sum_{k=1}^{\infty} z^k K_n'^* A_{+(n)}^{*(k-1)} C_{+(n)}^*$$

where $A_{+(n)}$ is the compression of A_+ to the first n components, $C_{+(n)}$ is the restriction of C_+ to the first n components and $K_n' = \alpha_n L_n^* \beta_n^* \beta_n'^*$ (α_n and β_n are those from Theorem A.4). It follows from the Appendix that

$$s - \lim_{n \to \infty} A_{+(n)}^* = A_+^*$$

and

$$s - \lim_{n \to \infty} C_{+(n)}^* = C_+^*.$$

Consequently, we have to prove only that

$$s \lim_{n \to \infty} K_n'^* P_{\mathcal{K}_n \ominus \mathcal{H}}^{\mathcal{K}_+ \ominus \mathcal{H}} = K_\infty^*$$

where $\mathcal{K}_n$ is given by (2.36). For this purpose, note first that $\mathcal{K}_+ \ominus \mathcal{H} = \overline{\bigcup_{n=1}^{\infty} D_n \mathcal{K}_n}$. Consequently, for $k \in \bigcup_{n=1}^{\infty} D_n \mathcal{K}_n$, there exists some $n_o \geq 1$ with $k = D_{n_o} k_{n_o}, k_{n_o} \in \mathcal{K}_{n_o}$. For $n_o \leq p < \infty$, we have

$$k = D_p k_{n_o}$$

(of course, k_{n_o} is naturally viewed also as an element in $\mathcal{K}_p$) and

$$L_p k_{n_o} = L_{n_o} k_{n_o}.$$

It follows that

$$\begin{aligned} K_\infty^* k &= \beta(L) L \alpha(L)^* k = \beta(L) L \alpha(L)^* D_\infty k_{n_o} \\ &= \beta(L) L D_L k_{n_o} = \beta(L) D_{L^*} L k_{n_o} = H_\infty(L) L k_{n_o}. \end{aligned}$$

For $p \geq n_o$,

$$\begin{aligned} K_p'^* k &= \beta_p' \beta_p L_p \alpha_p^* k = \beta_p' \beta_p L_p \alpha_p^* D_p k_{n_o} \\ &= \beta_p' \beta_p L_p D_{L_p} k_{n_o} = \beta_p' \beta_p D_{L_p^*} L_p k_{n_o} \\ &= \beta_p' H_p L_p k_{n_o} = H_p' L_p k_{n_o}. \end{aligned}$$

But now we use Lemma A.5 and we can conclude the proof that $\{\beta_n' \tilde{v}_n\}_{n \geq 1}$ converges coefficient–wise in the strong operatorial topology to R_F.

For the other part of the theorem use

$$I - a_n^*(e^{it}) a_n(e^{it}) = c_n^*(e^{it}) c_n(e^{it}) \quad a.e. \text{on } \mathbf{T}$$

and, as c_n is outer, $L_{a_n} = c_n$. The rest is as above. ∎

We particularise now to the scalar case. Let $F \in \mathcal{S}$ and $\{g_n\}_{n=0}^{\infty}$ be its Schur parameters (suppose also that $| g_n |< 1$ for $n \geq 0$). Schur's algorithm produces the polynomials $\{\mathcal{A}_n\}$ and $\{\mathcal{B}_n\}$ associated to F by formula (1.2). Moreover, the operatorial version in Section 1.3 produces four sequences of functions; $\{a_n\}$, $\{b_n\}$, $\{c_n\}$, $\{d_n\}$ by formula (1.39). The next result establishes the connections between all these functions.

Lemma 3.20 For $n \geq 0$,

$$\begin{aligned} a_n &= \frac{\mathcal{A}_n}{\mathcal{B}_n}, \\ b_n &= (\frac{z^n}{\mathcal{B}_n}) \prod_{k=0}^{n} (1- | g_k |^2)^{\frac{1}{2}}, \\ c_n &= (\frac{1}{\mathcal{B}_n}) \prod_{k=0}^{n} (1- | g_k |^2)^{\frac{1}{2}} \end{aligned}$$

and

$$d_n = -z(\frac{\mathcal{A}_n^T}{\mathcal{B}_n}).$$

Proof The proof follows immediately by comparing the formula (1.2) and the formula (1.38) (and taking into account the outerness of c_n and b_n/z^n). ∎

In this way the result in Theorem 3.19 yields a rational approximation of the spectral factor of F in the scalar case. One more remark is that

$$G_F^2(0) = \exp(\frac{1}{4\pi}\int_0^{2\pi} \log(1- \mid F(e^{it}) \mid^2)dt$$

by the formula in the Remark 3.5 and $G_F^2(0) = \prod_{n=0}^{\infty}(1- \mid g_n \mid^2)$ by the formula in Remark 2.4. Thus, we have the following result.

Corollary 3.21H Let $F \in \mathcal{S}$ and suppose that $\log(1- \mid F \mid^2) \in L^1$. Then, $(1/\mathcal{B}_n)\prod_{k=0}^{n}(1- \mid g_k \mid^2)^{\frac{1}{2}}$ converges uniformly on compact sets in $\mathbf{D}$ to g_F.

3.4 Some factorable functions

In view of the applications it is desirable to know large classes of functions for which the spectral factor realises an equality in its property (i) in Theorem 3.7 (or in (i′) at the beginning of Section 3.2. We call such functions factorable functions. The criterion in Proposition 3.8 is quite intricate and other criteria are to be taken into account.

Remark 3.22 Let a function N on $[0, 2\pi]$ be taken with values in $\mathcal{L}(\mathcal{H}), \mathcal{H}$ being a separable Hilbert space, and suppose that N is strongly measurable and $0 \leq N(t) \leq I$ for $t \in [0, 2\pi]$. Consider the semispectral measure

$$\mu(f) = \frac{1}{2\pi}\int_0^{2\pi} f(e^{it})N^2(t)dt, \quad f \in C(\mathbf{T})$$

and the operator

$$V : \mathcal{H} \to L^2(\mathcal{H})$$
$$(Vh)(t) = N(t)h.$$

Then it is obvious that

$$\mu(f) = V^* f(e^{it})V \quad \text{for } f \in C(\mathbf{T})$$

(Actually, $f(e^{it})$ is the functional calculus of the multiplication with e^{it} in $L^2(\mathcal{H})$). Consequently, the Naimark dilation of μ will be a part of e^{it} on $L^2(\mathcal{H})$; more precisely, with the notation in Theorem 3.7,

$$\mathcal{K}_- = \overline{NH^2(\mathcal{H})}$$

(where N is alternatively viewed as the multiplication with N on $L^2(\mathcal{H})$), and, in order to have equality in (i′) at the beginning of Section 3.2, it is necessary and sufficient to have

$$\bigcap_{n=0}^{\infty} e^{int}\overline{NH^2(\mathcal{H})} = 0.$$

Now we obtain a simpler criterion for factorability in the case when $\mathcal{H}$ is of finite dimension.

Proposition 3.23 Let $\mathcal{H}$ be a Hilbert space of finite dimension and $N : [0, 2\pi]$ a measurable function whose values are self–adjoint positive contractions on $\mathcal{H}$. Then if $\log\det N \in L^1$, N^2 is factorable.

Proof Let us define

$$m(t) = \inf\{(N(t)h, h) \mid h \in \mathcal{H}, \| h \| = 1\}.$$

Let $n = \dim \mathcal{H}$ and $\{\rho_k(t)\}_{k=1}^n$ the eigenvalues of $N(t)$ arranged in a non–decreasing order. Then

$$\det N(t) = \prod_{k=1}^{n} \rho_k(t)$$

and

$$m(t) = \rho_1(t).$$

As $\log\det N \in L^1$, $\rho_1(t) > 0$ a.e. and

$$n \log m(t) \leq \log\det N(t) \leq \log m(t) :$$

therefore $m \in L^1$.

By Theorem 3.2, there exists an outer function $G \in H^1$ with $m(t) = | G(e^{it}) |$ a.e. on $\mathbf{T}$. Then, we remark that for $F \in H^2(\mathcal{H})$,

$$\begin{aligned} \| NF \|^2 &= \frac{1}{2\pi} \int_0^{2\pi} \| N(t)F(e^{it}) \|^2 \, dt \\ &\geq \frac{1}{2\pi} \int_0^{2\pi} m(t)^2 \, \| F(e^{it}) \|^2 \, dt \\ &= \| GF \|^2 . \end{aligned}$$

In other words, we have obtained a contraction

$$X : \overline{NH^2(\mathcal{H})} \to H^2(\mathcal{H})$$
$$XNF = GF.$$

Finally, on the one hand,

$$X \bigcap_{n\geq 0} e^{int}\overline{NH^2(\mathcal{H})} \subset \bigcap_{n\geq 0} e^{int}\overline{XNH^2(\mathcal{H})} \subset \bigcap_{n\geq 0} e^{int}H^2(\mathcal{H}) = 0$$

and on the other hand X is injective because, for $F \in \overline{NH^2(\mathcal{H})}$ with $XF = 0$, there exist $F_n \in H^2(\mathcal{H})$ such that $GF_n = XNF_n \to XF = 0$ in $L^2(\mathcal{H})$. As $GN = NG$, it follows $GF = 0$, therefore $F = 0$. In this way, $\bigcap_{n\geq 0} e^{int}\overline{NH^2(\mathcal{H})} = 0$ and Theorem 3.7 and Remark 3.22 can be applied. ∎

The above argument is quite typical. Thus, let us define, for a separable Hilbert space $\mathcal{H}$ the class of functions

$$\mathcal{P}L^\infty(\mathcal{L}(\mathcal{H})) = \{N \in L^\infty(\mathcal{L}(\mathcal{H})) \mid \text{ there exists an inner scalar function } \mathcal{F} \text{ such that } F(e^{it})N^*(t) \in H^\infty(\mathcal{L}(\mathcal{H}))\}.$$

Proposition 3.24 If N if any positive function in $\mathcal{P}L^\infty(\mathcal{L}(\mathcal{H}))$, then N^2 is factorable.

Proof We can of course suppose $0 \leq N \leq I$. Let F be the scalar inner function for which $F(e^{it})N(t) \in H^\infty(\mathcal{L}(\mathcal{H}))$. Then,

$$\begin{aligned}\bigcap_{n\geq 0} e^{int}\overline{NH^2(\mathcal{H})} &= \overline{F(e^{it})} \bigcap_{n\geq 0} e^{int}\overline{F(e^{it})N(t)H^2(\mathcal{H})} \\ &\subset \overline{F(e^{it})} \bigcap_{n\geq 0} e^{int}H^2(\mathcal{H}) = 0\end{aligned}$$

and Theorem 3.7 and Remark 3.22 can be applied. ∎

Remark 3.25 The above argument can be used once more to show in both Propositions 3.23 and 3.24 that N is actually factorable. Indeed, as $N(t) \geq N^2(t)$ (because we can suppose $0 \leq N \leq I$), there exists a contraction

$$X : \overline{NH^2(\mathcal{H})} \to \overline{NH^2(\mathcal{H})}$$
$$XN^{\frac{1}{2}}F = NF, \quad F \in H^2(\mathcal{H}).$$

It is clear that X is injective and

$$X \bigcap_{n\geq 0} e^{int}\overline{N^{\frac{1}{2}}H^2(\mathcal{H})} \subset \bigcap_{n\geq 0} e^{int}\overline{XN^{\frac{1}{2}}H^2(\mathcal{H})} = \bigcap_{n\geq 0} e^{int}\overline{NH^2(\mathcal{H})} = 0.$$

■

Remark 3.26 We can add to Proposition 3.24 the following fact. Let F be the scalar inner function for which $F(e^{it})N(t) \in H^\infty(\mathcal{L}(\mathcal{H}))$ and let

$$N(t) = G(e^{it})^* G(e^{it})$$

be the factorisation produced by Proposition 3.24 and Remark 3.25, with G an outer function in $H^\infty(\mathcal{H}, \mathcal{G})$. Then we also have

$$F(e^{it})G(e^{it})^* \in H^\infty(\mathcal{G}, \mathcal{H}).$$

Indeed, $F(e^{it})N(t) \in H^\infty(\mathcal{L}(\mathcal{H}))$, i.e. $FNH^2(\mathcal{H}) \subset H^2(\mathcal{H})$, or $FG^*GH^2(\mathcal{H}) \subset H^2(\mathcal{H})$. But G is outer, so that $\overline{GH^2(\mathcal{H})} = H^2(\mathcal{G})$; consequently

$$FG^*H^2(\mathcal{G}) \subset H^2(\mathcal{H}),$$

which means that $FG^* \in H^\infty(\mathcal{G}, \mathcal{H})$.

One more class of factorable functions is given by the Schur algorithm (in its operatorial variant).

Proposition 3.27 If $F \in \mathcal{S}(\mathcal{H})$ has the Schur parameters $\{G_n\}_{n\geq 0}$ such that there exists a certain $N \in \mathbf{N}$ with $G_n = 0$ for $n > N$, then $I - F^*F$ and $I - FF^*$ are factorable.

Proof In view of the results in Section 1.3, we find that $F = a_N$, and the fact that the block matrix

$$\begin{bmatrix} a_N & b_N \\ c_N & d_N \end{bmatrix}$$

given by (1.37) is inner from both sides, gives the required factorisations.

■

We can obtain now an operatorial variant of Theorem 3.1.

Corollary 3.28 Any trigonometric polynomial $P(e^{it}) = \sum_{k=0}^{n}(A_k e^{-ikt} + A_k e^{ikt})$, $A_n \neq 0$, which is positive, admits the factorisation $P(e^{it}) = Q(e^{it})^* Q(e^{it})$, where $Q(z) = \sum_{k=0}^{n} B_k z^k$ is an analytic trigonometric polynomial and $B_k \in \mathcal{L}(\mathcal{H})$, $k = 0, \cdots, n$.

Proof It is obvious that $\chi_n P \in H^\infty(\mathcal{L}(\mathcal{H}))$ (recall that $\chi_n(z) = z^n$ for $n \geq 0$), so that $P \in \mathcal{P}L^\infty(\mathcal{L}(\mathcal{H}))$ and by Proposition 3.24 and Remark 3.26,

$$P(e^{it}) = Q(e^{it})^* Q(e^{it})$$

with Q an outer function. By Remark 3.26, $\chi_n Q^* \in H^\infty(\mathcal{G}, \mathcal{H})$; consequently Q has the required form (Q being outer, $\dim \mathcal{H} \geq \dim \mathcal{G}$ and B can be viewed as operators in $\mathcal{L}(\mathcal{H})$). ∎

Remark 3.29 There is a subclass of $\mathcal{P}L^\infty(\mathcal{L}(\mathcal{H}))$ which is especially interesting for (engineering) applications. This class consists in functions R with the representation $R = P/d$, where P is a polynomial with coefficients in $\mathcal{L}(\mathcal{H})$ and d is a polynomial with no zero in $\mathbf{D}$. As $d(z) = \prod_{k=1}^{n}(z - z_k)$, $| z_k | > 1$, we have

$$R(e^{it})^* = (-1)^n e^{int} P(e^{it})^* / \bar{z}_1 \cdots \bar{z}_n \prod_{k=1}^{n}(e^{it} - (1/\bar{z}_k))$$

and considering b to be the finite Blaschke product with zeros $(1/\bar{z}_1), \cdots, (1/\bar{z}_n) \in \mathbf{D}$, we get

$$e^{iNt} b(e^{it}) R(e^{it})^* \in H^\infty(\mathcal{L}(\mathcal{H}))$$

for sufficiently large N.

Notes

The problem of factorisation of positive functions goes back to the work of Fejér and Riesz (see [61] for a detailed discussion) for the case of trigonometric polynomials (Theorem 3.1), and in view of the physical and mathematical applications, much work was done to extend this result.

The extension to Hardy spaces was obtained by Szegö, together with several applications in the theory of orthogonal polynomials on the unit circle.

Since the importance of this type of factorisation for matrix (operator) valued functions was recognised (for instance, see [86] and [90] for applications in prediction theory of multivariate stationary processes), many generalisations appeared in this direction. Probably the geometric method of Lowdenslager using the Wold decomposition is the most fruitful in this area, leading to more and more general results, culminating in the work of Helson and Lowdenslager [51], de Branges [16], Rosenblum and Rovnyak [74] and [75] and Sz. Nagy and Foiaş [84]. For semispectral measures on the unit circle (i.e. Theorem 3.7), the result was proved in [79].

Remark 3.11 concerning the connection between this type of factorisation, Cholesky factorisation and factorisation of positive definite kernels on the set of integers is taken from [28]. Proposition 3.13, Theorem 3.15, Corollary 3.16 and Remark 3.17 and 3.18 are in [26]. Theorem 3.19 was proved in [11] extending some results in Szegö's theory of orthogonal polynomials from the scalar case; see [14], [48] and [83].

Classes of factorable functions are of a wide interest. Proposition 3.23 has a certain history; see [75] or [84]. Proposition 3.24 is a particular case of a result proved in [74] and for Corollary 3.28 and Remark 3.29 see also [74]. A few other papers in this field are: [32], [37]. [87] and [88].

CHAPTER 4

Orthogonal polynomials on the unit circle

Strongly related to Schur's algorithm are the recurrence formulas satisfied by the orthogonal polynomials on the unit circle. The purpose of this chapter is to make explicit this connection and connections with dilation theory will also appear. We take into consideration the algebraic and asymptotic aspects of Szegö's theory of orthogonal polynomials on the unit circle. Most of this theory is of a geometric nature and in Section 4.3 we briefly describe some recent developments beyond Szegö's theory. Additionally, in the last section, an operatorial version of Szegö's theory is presented.

4.1 Recurrence formulas

We consider a finite positive Borel measure μ on $\mathbf{T}$ and we let $L^2(\mu)$ be the space of measurable functions on $\mathbf{T}$ of integrable square with respect to μ; the scalar product is given by:

$$(f,g) = \frac{1}{2\pi}\int_0^{2\pi} f(e^{it})\overline{g(e^{it})}d\mu(t)$$

for $f, g \in L^2(\mu)$. Let us consider the functions $\chi_n(z) = z^n$, $n \geq 0$, and let $\{s_n\}_{n=-\infty}^{\infty}$ be the Fourier coefficients of μ. Moreover, we have defined for $n \geq 0$ the matrices

$$M_n = \begin{bmatrix} s_0 & s_1 & \cdots & s_n \\ \bar{s}_1 & s_o & \cdots & s_{n-1} \\ \bar{s}_n & \bar{s}_{n-1} & \cdots & s_o \end{bmatrix}.$$

M_n is the grammian of the family $\{\chi_p\}_{p=0}^n$, so that $\{\chi_p\}_{p=0}^n$ is a linearly independent family in $L^2(\mu)$ if and only if $\det M_n \neq 0$.

Let us consider such a linearly independent family $\{\chi_p\}_{p=0}^n$ generaring in $L^2(\mu)$ a finite dimensional space $\mathcal{P}_n$, the space of polynomials of degree at most n. We choose in P_n the basis obtained by the Gramm–Schmidt procedure applied to the family $\{\chi_p\}_{p=0}^n$. In this way a family of polynomials (called Szegö Polynomials) $\{\phi_p\}_{p=0}^n$ is obtained such that:

(a) ϕ_p is a polynomial of degree p,

$$\phi_p(z) = \phi_p(\mu, z) = e_{pp}z^p + e_{p,p-1}z^{p-1} + \cdots + e_{po}, \quad e_{pp} > 0.$$

(b) $(\phi_p, \phi_m) = \delta_{pm}, \quad 0 \le p, m \le n.$

As e_{pp} plays a privileged role, we denote it by k_p. If we assume that $\det M_n \neq 0$ for all $n \geq 0$, we obtain a sequence of polynomials $\{\phi_p\}_{p=0}^{\infty}$ uniquely determined by the conditions (a) and (b) for all $p \geq 0$. Moreover, we define the monic orthogonal polynomials:

$$\Phi_p(z) = \Phi_p(\mu, z) = k_p^{-1}\phi_p(z), \quad p \geq 0.$$

The orthogonal polynomials satisfy a basic recurrence formula.

Theorem 4.1 The monic orthogonal polynomials satisfy the following recurrence relations.

(1) $\Phi_{p+1}(z) = z\Phi_p(z) + \Phi_{p+1}(0)\Phi_p^T(z).$

(2) $k_p\phi_{p+1}(z) = k_{p+1}z\phi_p(z) + \phi_{p+1}(0)\phi_p^T(z).$

Proof As the last coefficient of Φ_p^T is 1, it follows that

$$\frac{\Phi_{p+1}^T(z) - \Phi_p^T(z)}{z}$$

is a polynomial of degree p and for $0 \le m \le p-1$ we have:

$$\frac{1}{2\pi}\int_0^{2\pi} e^{imt}\overline{\left(\frac{\Phi_{p+1}^T(e^{it}) - \Phi_p^T(e^{it})}{e^{it}}\right)}d\mu(t)$$
$$= \frac{1}{2\pi}\int_0^{2\pi} e^{i(m-p)t}\Phi_{p+1}(e^{it})d\mu(t) - \frac{1}{2\pi}\int_0^{2\pi} e^{i(m-p+1)t}\Phi_p(e^{it})d\mu(t) = 0.$$

Consequently,

$$\frac{\Phi_{p+1}^T(z) - \Phi_p^T(z)}{z} = \alpha_p\Phi_p(z)$$

or

$$\Phi_{p+1}^T(z) = \Phi_p^T(z) + z\alpha_p\Phi_p(z).$$

Finally,

$$\Phi_{p+1}(z) = z\Phi_p(z) + \bar{\alpha}_p\Phi_p^T(z)$$

and $\bar{\alpha}_p = \Phi_{p+1}(0)$, so, part (1) is proved. Then (2) follows from (1). ∎

The numbers $\Phi_p(0)$ are usually called the Szegö parameters of μ.

Remark 4.2 Let us note an explicit formula for the orthogonal polynomials. From the orthogonality condition we have $(\phi_n, \chi_p) = 0$ for $0 \leq p \leq n-1$ and this can be rewritten as the system:

$$\sum_{p=0}^{n} e_{np} s_{m-p} = 0, \quad 0 \leq m \leq n-1$$

$$\sum_{p=0}^{n} e_{np} \chi_p = \phi_n \tag{4.1}$$

verified by the coefficients of ϕ_n. By the Cramer rule,

$$k_n = \frac{\phi_n \det M_{n-1}}{det \begin{bmatrix} s_o & \bar{s}_1 & \cdots & \bar{s}_n \\ s_1 & s_o & \cdots & \bar{s}_{n-1} \\ \cdot & & & \cdot \\ \cdot & & & \cdot \\ \cdot & & & \cdot \\ s_{n-1} & s_{n-2} & \cdots & \bar{s}_1 \\ \chi_o & \chi_1 & \cdots & \chi_n \end{bmatrix}}$$

and

$$\phi_n = \frac{k_n}{\det M_{n-1}} \det \begin{bmatrix} s_o & \bar{s}_1 & \cdots & \bar{s}_n \\ s_1 & s_o & \cdots & \bar{s}_{n-1} \\ \cdot & & & \cdot \\ \cdot & & & \cdot \\ \cdot & & & \cdot \\ s_{n-1} & s_{n-2} & \cdots & \bar{s}_1 \\ \chi_o & \chi_1 & \cdots & \chi_n \end{bmatrix}. \tag{4.2}$$

Now we can deduce a representation of the leading coefficient k_n in terms of the Fourier coefficients of μ. It is clear that

$$(\det \begin{bmatrix} s_o & \bar{s}_1 & \cdots & \bar{s}_n \\ s_1 & s_o & \cdots & \bar{s}_{n-1} \\ \cdot & & & \cdot \\ \cdot & & & \cdot \\ \cdot & & & \cdot \\ s_{n-1} & s_{n-2} & \cdots & \bar{s}_1 \\ \chi_o & \chi_1 & \cdots & \chi_n \end{bmatrix}, \chi_n) = \det M_n$$

and, on the other hand, as $\{\phi_p\}_{p=0}^{n}$ is a basis in $\mathcal{P}_n$, we deduce that

$$\chi_n = \frac{1}{k_n}\phi_n + \cdots.$$

Taking (4.2) into account, we get:

$$(\frac{\det M_{n-1}}{k_n}\phi_n, \frac{1}{k_n}\phi_n + \cdots) = \det M_n$$

or

$$\frac{1}{k_n^2} = \frac{\det M_n}{\det M_{n-1}}.$$

■

Remark 4.3 We see also from the system (4.1) that the coefficients of ϕ_p satisfy the equation:

$$M_p(c_{po}, \cdots, e_{p,p-1}, k_p)^{\mathrm{t}} = (0, \cdots, 0, \frac{1}{k_p})^{\mathrm{t}}. \tag{4.4}$$

As M_p is Toeplitz, from (4.4) we also deduce that

$$M_p(k_p, \bar{e}_{p,p-1}, \cdots, \bar{e}_{po})^{\mathrm{t}} = (\frac{1}{k_p}, 0, \ldots, 0)^{\mathrm{t}} \tag{4.5}$$

and the recurrence relations in Theorem 4.1 ensure that if $(e_{po}, \cdots, e_{p,p-1}, k_p)^{\mathrm{t}}$ satisfies (4.4) then

$$\begin{aligned} M_{p+1}(\frac{k_{p+1}}{k_p}&(0, e_{po}, \cdots, e_{p,p-1}, k_p)^{\mathrm{t}} \\ &+ \frac{e_{p+1,o}}{k_p}(k_p, \bar{e}_{p,p-1}, \cdots, \bar{e}_{po}, 0)^{\mathrm{t}}) \\ &= (0, \cdots, 0, \frac{1}{k_{p+1}})^{\mathrm{t}}. \end{aligned} \tag{4.6}$$

We define

$$\Delta_p = s_1 e_{po} + \cdots + s_p e_{p,p-1} + s_{p+1} k_p$$

and from (4.4), (4.5) and (4.6) it follows that

$$\frac{k_{p+1}}{k_p}\Delta_p + \frac{e_{p+1,o}}{k_p^2} = 0 \tag{4.7}$$

and

$$\frac{k_{p+1}}{k_p^2} + \frac{e_{p+1,o}}{k_p}\bar{\Delta}_p = \frac{1}{k_{p+1}}. \tag{4.8}$$

First, we can rewrite (4.7) in order to get

$$\Delta_p = -\frac{\phi_{p+1}(0)}{k_p k_{p+1}} = -\frac{\Phi_{p+1}(0)}{k_p}$$

and then (4.8) becomes:

$$\frac{1}{k_{p+1}^2} = \frac{1}{k_p^2}(1- \mid \Phi_{p+1}(0) \mid^2).$$

From (4.9) we deduce a representation of the leading coefficient k_n in terms of the Szegö parameters:

$$\frac{1}{k_n^2} = s_o \prod_{p=1}^{n}(1- \mid \Phi_p(0) \mid^2). \tag{4.10}$$

Moreover, from (4.3) and (4.10), we can obtain a formula for computing the determinant of M_n:

$$\det M_n = s_0^{n+1} \prod_{p=1}^{n}(1- \mid \Phi_p(0) \mid^2)^{n-p+1} \tag{4.11}$$

which shows that $\det M_n \neq 0$ if and only if $\mid \Phi_o(0) \mid < 1$ for all $p, 0 \leq p \leq n$, or, that the family $\{\chi_p\}_{p=0}^n$ is linearly independent if and only if $\mid \Phi_p(0) \mid < 1$, $0 \leq p \leq n$. ■

Another consequence of the recurrence formulas is the following Christoffel–Darboux–type formula.

Theorem 4.4 for $z\bar{w} \neq 1$, the following formula holds:

$$\sum_{p=0}^{n-1} \phi_p(z)\overline{\phi_p(w)} = \frac{\phi_n^T(z)\overline{\phi_n^T(w)} - \phi_n(z)\overline{\phi_n(w)}}{1 - z\bar{w}}.$$

Proof By Theorem 4.1 and (4.9) we get for $0 \leq p \leq n$:

$$\begin{aligned} &\phi_p^{\mathrm{T}}(z)\overline{\phi_p^{\mathrm{T}}(w)} - \phi_p(z)\overline{\phi_p(w)} \\ &\quad = \phi_{p-1}^{\mathrm{T}}(z)\overline{\phi_{p-1}^{\mathrm{T}}(w)} - z\bar{w}\phi_{p-1}(z)\overline{\phi_{p-1}(w)}. \end{aligned} \tag{4.12}$$

Adding these equalities for $p = 0, \cdots, n$ we obtain exactly the required formula. ■

Now let us note some consequences of the Christoffel–Darboux formula.

Corollary 4.5 $\mid \phi_n^{\mathrm{T}}(z) \mid \geq \mid \phi_n(z) \mid$ for $z \in \bar{\mathbf{D}}$ and $\mid \phi_n^{\mathrm{T}}(z) \mid > \mid \phi_n(z) \mid$ for $z \in \mathbf{D}$.

Proof By Theorem 4.4 with $z = w$,

$$0 \leq \sum_{p=o}^{n-1} \mid \phi_p(z) \mid^2 = \frac{\mid \phi_n^{\mathrm{T}}(z) \mid^2 - \mid \phi_n(z) \mid^2}{1- \mid z \mid^2} \tag{4.13}$$

so that, for $z \in \mathbf{D}$, $|\phi_n^{\mathrm{T}}(z)| \geq |\phi_n(z)|$ and for $z \in \mathbf{T}$, $|\phi_n^{\mathrm{T}}(z)| = |\phi_n(z)|$. For the second inequality, if we have $|\phi_n^{\mathrm{T}}(z_o)| = |\phi_n(z_o)|$ for a certain $z_o \in \mathbf{D}$, it follows by (4.13) that $\phi_p(z_o) = 0$, $0 \leq p \leq n-1$, which contradicts $\phi_o(z_o) = s_o \neq 0$. ∎

Corollary 4.6 The roots of ϕ_n are in $\mathbf{D}$.

Proof By (4.13),

$$\begin{aligned} 0 < |\phi_o(z)|^2 \leq \sum_{p=o}^{n-1} |\phi_p(z)|^2 &= \frac{|\phi_n^{\mathrm{T}}(z)|^2 - |\phi_n(z)|^2}{1 - |z|^2} \\ &\leq \frac{|\phi_n^{\mathrm{T}}(z)|^2}{1 - |z|^2} \quad \text{for } z \in \mathbf{D}. \end{aligned}$$

Consequently, for $z \in \mathbf{D}$, $|\phi_n^{\mathrm{T}}(z)|^2 \geq |\phi_o(z)|^2 (1 - |z|^2) > 0$. Furthermore, if we have $\phi_n^{\mathrm{T}}(z_o) = 0 = \phi_n(z_o)$ for a certain $z_o \in \mathbf{T}$, it follows by Theorem 4.1 that $|\Phi_n(0)| = 1$, which contradicts Remark 4.3. ∎

Another important consequence of Theorem 4.4 is that we can construct on $\mathcal{P}_n$ a certain reproducing kernel. We say that a Hilbert space $\mathcal{H}$ of functions $f : X \to \mathbf{C}$ has a reproducing kernel if there exists an application

$$K : X \times X \to \mathbf{C}$$

such that:

(a) $K_y \in \mathcal{H}$, for every $y \in X$, where $K_y(x) = K(x, y)$.

(b) $f(y) = (f, K_y)$ for every $y \in X$ and $f \in \mathcal{H}$.

From these properties it follows that

$$K(y, x) = K_x(y) = (K_x, K_y)$$

which shows that if it exists, the reproducing kernel is unique and, moreover, it is positive definite: for every $x_1, \cdots, x_n$ in X and $z_1, \cdots, z_n$ in $\mathbf{C}$,

$$\sum_{i,j=1}^{n} K(x_i, x_j) z_j \bar{z}_i \geq 0.$$

Proposition 4.7

$$K^{(n)}(z,w) = \frac{\phi_{n+1}^{\mathrm{T}}(z)\overline{\phi_{n+1}^{\mathrm{T}}(w)} - \phi_{n+1}(z)\overline{\phi_{n+1}(w)}}{1 - z\bar{w}}$$

is a reproducing kernel for $\mathcal{P}_n$.

Proof That $K_w^{(n)}$ belongs to $\mathcal{P}_n$ is obvious from the Christoffel–Darboux formula. For $P \in \mathcal{P}_n$ we can compute, based on the same formula, that

$$(P, K_w^{(n)}) = (P, \sum_{p=0}^{n} \overline{\phi_p(w)}\phi_p) = \sum_{p=0}^{n} \phi_p(w)(P, \phi_p) = P(w). \qquad \blacksquare$$

4.2 Schur polynomials and Szegö polynomials

If we start with a positive measure μ on $\mathbf{T}$ with $\mu(1) = 1$, let $\{g_n\}_{n\geq 1}$ be its choice parameters given by Theorem 2.14 (i.e. the choice parameters of the matrices M_n, $n \geq 1$, associated to the Fourier coefficients of μ). We have also associated to μ, by the rules from Section 1.2, the functions:

$$G(z) = \frac{1}{2\pi}\int_0^{2\pi} \frac{e^{it}+z}{e^{it}-z} d\mu(t) \in \mathcal{C} \tag{4.14}$$

and

$$F(z) = \frac{G(z)-1}{z(G(z)+1)} \in \mathcal{S}. \tag{4.15}$$

The Schur parameters of F are exactly $\{g_n\}_{n\geq 1}$. Moreover, we have the Schur polynomials $\{\mathcal{A}_n\}_{n\geq 0}$, $\{\mathcal{B}_n\}_{n\geq 0}$ associated to F in Section 1.1 and the polynomials $\{p_n\}_{n\geq 1}$ given by (2.49). In the previous section we defined the Szegö polynomials of μ. It is necessary, of course, to know the connection between all these polynomials.

Theorem 4.8 $\bar{p}_n = \Phi_n$ for all $n \geq 1$.

Proof We will show that $\{p_n\}_{n\geq 1}$ satisfy $(\bar{p}_n, \chi_p) = 0$, $0 \leq p < n$, and, as Φ_n and $\bar{p}_n$ are monic polynomials, they will coincide. Recall now the realisation of $L^2(\mu)$ established in Remark 2.26. First of all there is a unitary operator denoted by ω identifying $L^2(\mu)$ with $\ell^2(\mathbf{Z}, \mathcal{T}$, where $\mathcal{T}$ is the positive definite Toeplitz kernel built on the Fourier coefficients of μ. This operator ω is defined by $\omega(\chi_{-n}) = e_n$, $n \in \mathbf{Z}$, where $e_n = (, \cdots, 0, 1, 0, \cdots)$ with 1 on the nth entry. Let $W = W(\{g_n\}_{n=1}^{\infty})$ be the

Naimark dilation of μ acting on the space $\mathcal{K} = \cdots \bigoplus \mathcal{D}_*(L) \bigoplus \mathbf{C} \bigoplus \mathbf{C} \bigoplus \cdots$, with $L = L(\{g_n\}_{n=1}^{\infty})$. Taking into account the operator Ω defined by (2.45), $\Omega\omega$ is a unitary operator from $L^2(\mu)$ into $\mathcal{K}$ defined by $\Omega\omega(\chi_{-n}) = W^{*n}e_o$ for $n \in \mathbf{Z}$ Moreover, the operator $\Omega\omega$ yields a unitary operator when restricted from $\mathcal{P}_n$ onto $\mathcal{K}_{n+1}$ ($= \mathbf{C}^{n+1}$, in our scalar case).

Having the specified action of $\Omega\omega$, we get

$$\Omega\omega\bar{p}_n = \bar{p}_n(W^*)e_o = (\cdots *, *, \cdots, \bar{p}_n(W_n^*)e_o, *, \cdots)^t$$

where $\bar{p}_n(W_n^*)e_o$ occurs on the positions indexed by $0, 1, \cdots, n$ (and W_n is that given by (2.36). The entries marked by '$*$' play no role here, but,

$$\bar{p}_n(z) = \overline{p_n(\bar{z})} = \overline{\det(\bar{z}I - W_n)} = \det(zI - W_n^*).$$

so that, by the Cayley–Hamilton theorem, $\bar{p}_n(W_n^*) = 0$, and for $0 \leq k < n$,

$$(\bar{p}_n, \chi_k) = (\Omega\omega\bar{p}_n, \Omega\omega\chi_k) = 0. \qquad \blacksquare$$

Remark 4.9 Now we can establish the connection between the Szegö parameters of μ and the Schur parameters of the function F in the class $\mathcal{S}$ associated to μ, namely

$$\Phi_n(0) = -\bar{g}_n, \quad n \geq 1.$$

Indeed, by Theorem 4.8,

$$\Phi_n(0) = \bar{p}_n(0).$$

By definition of p_n,

$$\bar{p}_n(0) = \det(-W_n^*)$$

and by the definition (2.37) of W_n,

$$\begin{aligned} \det(-W_n^*) &= (-1)^n \det(I \bigoplus \bar{g}_n) \det V^*(\{g_k\}_{k=1}^{n-1}) \\ &= (-1)^n \bar{g}_n \det V^*(\{g_k\}_{k=1}^{n-1}). \end{aligned}$$

But now, $\det V^*(\{g_k\}_{k=1}^{n-1}) = (-1)^{n-1}$ because

$$\det \begin{bmatrix} g_k & d_k \\ d_k & -\bar{g}_k \end{bmatrix} = -1$$

and finally,

$$\Phi_n(0) = -\bar{g}_n. \qquad \blacksquare$$

The last result of this section contains the connection between Szegö polynomials and Schur polynomials.

Theorem 4.10 For $n \geq 1$

$$\Phi_n(z) = z\mathcal{B}^{\mathrm{T}}_{n-1}(z) - \mathcal{A}^{\mathrm{T}}_{n-1}(z).$$

Moreover, $z\mathcal{B}^{\mathrm{T}}_{n-1}(z) + \mathcal{A}^{\mathrm{T}}_{n-1}(z)$ are the monic orthogonal polynomials of the measure determined by the choice parameters $\{-g_n\}_{n=1}^{\infty}$.

Proof By (1.5), Proposition 2.2 and Remark 4.9,

$$\begin{aligned} z(\mathcal{B}^{\mathrm{T}}_n(z) - \mathcal{A}^{\mathrm{T}}_n(z) &= (z\mathcal{B}^{\mathrm{T}}_{n-1}(z) - \mathcal{A}^{\mathrm{T}}_{n-1}(z)) - \bar{g}_{n+1}(\mathcal{B}_{n+1}(z) - z\mathcal{A}_{n-1}(z)) \\ &= z\mathcal{B}^{\mathrm{T}}_{n-1}(z) - \mathcal{A}^{\mathrm{T}}_{n-1}(z) + \Phi_{n+1}(0)(\mathcal{B}_{n-1}(z) - z\mathcal{A}_{n-1}(z)). \end{aligned}$$

On the other hand, by (1.4), $\mathcal{B}_n(0) = 1$ for $n \geq 0$, so $\mathcal{A}^{\mathrm{T}}_n(0) = \bar{g}_{n+1}$ for $n \geq 0$. Consequently,

$$(z\mathcal{B}^{\mathrm{T}}_n(z) - \mathcal{A}^{\mathrm{T}}_n(z))(0) = -\mathcal{A}^{\mathrm{T}}_n(0) = -\bar{g}_{n+1} = \Phi_{n+1}(0)$$

and

$$\Phi_n(z) = z\mathcal{B}^{\mathrm{T}}_{n-1}(z) - \mathcal{A}^{\mathrm{T}}_{n-1}(z).$$

Similar computations show that the polynomials

$$\Psi_n(z) = z\mathcal{B}^{\mathrm{T}}_{n-1}(z) + \mathcal{A}^{\mathrm{T}}_{n-1}(z), \quad n \geq 1,$$

of degree n, satisfy the relations:

$$\Psi_{n+1}(z) = z\Psi_n(z) + \bar{g}_{n+1}\Psi^{\mathrm{T}}_n(z)$$

and $\Psi_n(0) = \mathcal{A}^{\mathrm{T}}_{n+1}(0) = \bar{g}_n$. Thus, $\{\Psi_n\}_{n\geq 1}$ are the monic orthogonal polynomials determined by the choice parameters $\{-g_n\}_{n\geq 1}$. $\blacksquare$

Remark 4.11 The monic polynomials $\{\Psi_n\}_{n\geq 0}$ ($\Psi_o(z) = 1$) are usually called Geronimus polynomials or orthogonal polynomials of the second kind. It is a simple computation to show that they are obtained by dividing the Szegö polynomials $\{\psi_n\}_{n\geq 0}$ corresponding to the parameters $\{-g_n\}_{n\geq 1}$ by their highest index coefficients. Thus they admit the alternative description:

$$\psi_o(z) = 1$$
$$\psi_n(z) = \frac{1}{2\pi}\int_0^{2\pi} \frac{e^{it}+z}{e^{it}-z}(\phi_n(e^{it}) - \phi_n(z))d\mu(t) \quad \text{for } n \geq 1.$$

Remark 4.12 Most of the connections between the objects involved in our considerations can be summarised now in the following diagram.

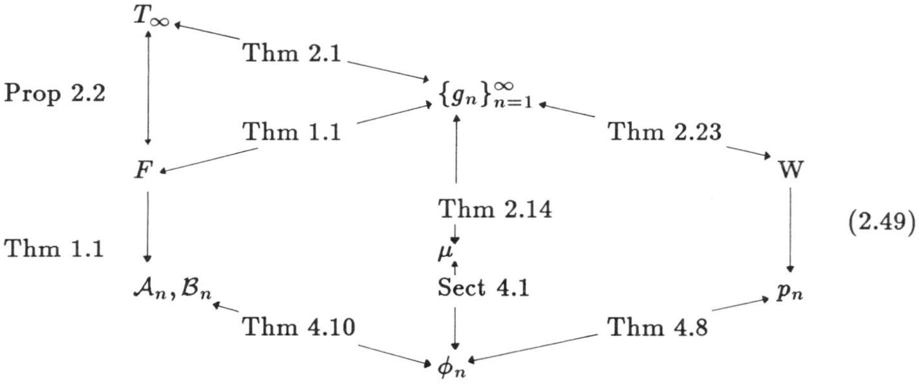

(2.49)

Remark 4.13 It is now the time to note that the polynomials $\hat{\Phi}_n$ and $\hat{\Psi}_n$ which appeared in Theoren 1.7 are exactly the Szegö polynomials and Geronimus polynomials of the measure μ having as Fourier coefficients the elements of M_n.

4.3 Szegö's theory and generalisations

Szegö's theory concerns the asymptotic properties of the orthogonal polynomials in the case when the absolutely continuous part $d\mu/dt = \mu'$ of the given measure μ satisfies Szegö's condition

$$\int_0^{2\pi} \log \mu'(t)dt > -\infty \tag{4.16}$$

(i.e. $\log \mu' \in L^1$). Recall that this is also the condition for the existence of the spectral factor G_μ of μ (also called the Szegö function of μ).

The main result of Szegö's theory is the following.

Theorem 4.14 Let μ be a positive measure on $\mathbf{T}$ satisfying Szegö's condition (4.16). Then

(1) $\lim_{n\to\infty} \phi_n^{\mathrm{T}}(z) = G_{\mu}^{-1}(z)$

(2) $\lim_{n\to\infty} \phi_n(z) = 0$

(3) $\sum_{n=0}^{\infty} \phi_n(z)\overline{\phi_n(w)} = (1 - z\bar{w})^{-1} G_{\mu}^{-1}(z)\overline{G_{\mu}^{-1}(w)}$

where the convergence in (1)–(3) is uniform on compact subsets of $\mathbf{D}$.

Proof Let F be the function in $\mathcal{S}$ associated to μ by (4.15), and let $\{\mathcal{A}_n\}_{n\geq 0}$, $\{\mathcal{B}_n\}_{n\geq 0}$ be the Schur polynomials of F. Then, by Theorem 4.10

$$\begin{aligned}\phi_n^{\mathrm{T}}(z) &= k_n \Phi_n^{\mathrm{T}}(z) = k_n(\mathcal{B}_{n-1}(z) - z\mathcal{A}_{n-1}(z)) \\ &= k_n \mathcal{B}_{n-1}(z)(1 - z\frac{\mathcal{A}_{n-1}(z)}{\mathcal{B}_{n-1}(z)}).\end{aligned}$$

By Theorem 1.1, $\{\mathcal{A}_n/\mathcal{B}_n\}_{n\geq 0}$ converges uniformly on compact subsets of $\mathbf{D}$ to F and by Corollary 3.21, $\{k_n\mathcal{B}_n\}_{n\geq 0}$ converges uniformly on compact subsets of $\mathbf{D}$ to G_F^{-1}, G_F being the spectral factor of F. (It is worthwhile recalling that the Schur parameters of F are $\{g_n\}_{n\geq 1}$.) Consequently,

$$\lim_{n\to\infty} \phi_n^{\mathrm{T}}(z) = \frac{1 - zF(z)}{G_F(z)} = G_{\mu}^{-1}(z)$$

uniformly on compact subsets of $\mathbf{D}$, where we used Remark 3.5 and thus (1) is proved.

We already know that $\mu' \in L^1$ if and only if

$$1/k_{\infty}^2 = \prod_{n=1}^{\infty}(1 - \mid g_n \mid^2) \tag{4.17}$$

is strictly positive. Consequently, by (1.7), $\mathcal{B}_n$ is bounded from below on $\mathbf{D}$. By Theorem 1.3

$$\frac{\mathcal{A}_n(z) + z\mathcal{B}_n^{\mathrm{T}}(z)}{\mathcal{B}_n(z) + z\mathcal{A}_n^{\mathrm{T}}(z)}$$

converges uniformly on compact subsets of $\mathbf{D}$ to F. Thus $\mathcal{A}_n^{\mathrm{T}}$ and $\mathcal{B}_n^{\mathrm{T}}$ converge uniformly on compact subsets of $\mathbf{D}$ to zero, and, as

$$\phi_n(z) = k_n(z\mathcal{B}_n^{\mathrm{T}}(z) - \mathcal{A}_n^{\mathrm{T}}(z))$$

by Theorem 4.10, (2) follows.

(3) follows from (1) and (2) and Theorem 4.4. ∎

Let us note some other results in Szegö's theory. First, an important extremal problem which has significance in prediction theory. Thus, for μ, consider $L^2(\mu)$. The closed space $\mathcal{P}$ generated by polynomials can be thought of as the past of the process $\{e^{int}\}_{n\in\mathbf{Z}}$ evoluting in $L^2(\mu)$. One problem is to compute the angle between the past and the present, i.e. the angle between $\mathcal{P}$ and e^{-it}. Moreover, it is interesting to compute the angle between $\mathcal{P}_n$ and e^{-it}. Thus, we have the following extremal problems.

'Compute

$$\alpha = \inf_{\substack{p\in\mathcal{P}\\ p(0)=1}} \frac{1}{2\pi}\int_0^{2\pi} |\, p(e^{it})\,|^2\, d\mu(t)$$

and

$$\alpha_n = \inf_{\substack{p\in\mathcal{P}_n\\ p(0)=1}} \frac{1}{2\pi}\int_0^{2\pi} |\, p(e^{it})\,|^2\, d\mu(t).'$$

Theorem 4.15

$$\alpha = \prod_{n=1}^{\infty}(1-|\, g_n\,|^2) = |\, G\mu(0)\,|^2$$

$$\alpha_n = \prod_{k=1}^{n}(1-|\, g_k\,|^2) = |\, \phi_n^{\mathrm{T}}(0)\,|^{-2}$$

$$= \frac{1}{\sum_{k=0}^{n} |\, \phi_k(0)\,|^2}.$$

Proof We have

$$\alpha_n = \inf_{\substack{a_o,\cdots,a_n\\ a_o=1}} \frac{1}{2\pi}\int_0^{2\pi} |\, p(e^{it})\,|^2\, d\mu(t)$$

$$
\begin{aligned}
&= \inf_{\substack{a_o,\cdots,a_n\\ a_o=1}} \sum_{k,j=0}^{n} \frac{1}{2\pi}\int_0^{2\pi} e^{i(k-j)t} a_k \bar{a}_j d\mu(t)\\
&= \inf_{\substack{a_o,\cdots,a_n\\ a_o=1}} \sum_{k,j=0}^{n} (a_k \tilde{W}^{*k} e_o, a_j \tilde{W}^{*j} e_o)\\
&= \inf_{a_1,\cdots,a_n} \| (I - \sum_{k=1}^{n} a_k \tilde{W}^{*k}) e_o \|^2\\
&= \| (I - P_n) e_o \|^2\\
&= (P_{\mathbf{C}}^{\tilde{\mathcal{K}}}(I - P_n) e_o, e_o),
\end{aligned}
$$

where $\tilde{W}$ is given by (3.17), $e_o = (\cdots, 0, 1, 0, \cdots) \in \tilde{\mathcal{K}}$ is the vector with 1 on position 0, and P_n is the orthogonal projection of $\tilde{\mathcal{K}}$ (the space where $\tilde{W}$ acts) onto $\bigvee_{k=1}^{n} \tilde{W}^{*k}\mathbf{C}$. Then, by a similar computation,

$$
\alpha = (P_{\mathbf{C}}^{\tilde{\mathcal{K}}}(I - P) e_o, e_o).
$$

But now,

$$
P_{\mathbf{C}}^{\tilde{\mathcal{K}}}(I - P_n) e_o = P_{\mathbf{C}}^{\tilde{\mathcal{K}}} \tilde{W}^*(I - P_{\tilde{\mathcal{K}}_{n-1}}^{\tilde{\mathcal{K}}}) \tilde{W} e_o
$$

and

$$
P_{\mathbf{C}}^{\tilde{\mathcal{K}}}(I - P) e_0 = P_{\mathbf{C}}^{\tilde{\mathcal{K}}} \tilde{W}^*(I - P_{\mathcal{K}_-}^{\tilde{\mathcal{K}}}) \tilde{W} e_o
$$

where $\mathcal{K}_-$ is defined by (3.18) and $\tilde{\mathcal{K}}_n = \bigvee_{k=0}^{n} \tilde{W}^{*k}\mathbf{C} = \mathbf{C}^{n+1}$ (in this scalar case, when, recall, the condition $| g_n | < 1$ is in force).

Finally

$$
\begin{aligned}
P_{\mathbf{C}}^{\tilde{\mathcal{K}}}(I - P_n) e_o &= e_o - P_{\mathbf{C}}^{\tilde{\mathcal{K}}} \tilde{W}^* P_{\tilde{\mathcal{K}}_{n-1}}^{\tilde{\mathcal{K}}} \tilde{W} e_0\\
&= (I - \tilde{L}(\{g_k\}_{k=1}^{n})^* \tilde{L}(\{g_k\}_{k=1}^{n})) e_0\\
&= \prod_{k=1}^{n} (1 - | g_k |^2) e_o
\end{aligned}
$$

and in a similar way

$$
P_{\mathbf{C}}^{\tilde{\mathcal{K}}}(I - P) e_o = \prod_{n=1}^{\infty} (1 - | g_n |^2) e_o.
$$

The rest is plain. ∎

The above result shows that the convergence in Theorem 4.14 (1) has a geometric nature, i.e. it reflects the convergence of certain angles in the process $\{e^{int}\}_{n\in\mathbf{Z}}$ evoluting in $L^2(\mu)$.

Another related problem is the following. For a positive measure μ on $\mathbf{T}$ define the 'entropy' of μ by the formula

$$h(\mu, z) = -\frac{1}{4\pi}\int_0^{2\pi} \mathrm{Re}(\frac{e^{it}+z}{e^{it}-z})\log\mu'(t)dt.$$

Then consider the complex numbers $s_1, \cdots, s_n$ such that

$$M_n = \begin{bmatrix} 1 & s_1 & \cdots & s_n \\ \bar{s}_1 & 1 & \cdots & s_{n-1} \\ \cdot & & & \cdot \\ \cdot & & & \cdot \\ \cdot & & & \cdot \\ \bar{s}_n & \bar{s}_{n-1} & \cdots & 1 \end{bmatrix}$$

is a positive Toeplitz matrix. The problem is the following.

'Compute

$$h(z) = \min\{h(\mu, z) \mid \mu \text{ is a positive measure with the first } n+1 \text{ Fourier coefficients } 1, s_1, \cdots, s_n\}.'$$

Proposition 4.16

$$h(z) = \log(1- \mid z \mid^2)^{\frac{1}{2}} + \frac{1}{2}\log\sum_{k=0}^{n} \mid \phi_k(z) \mid^2$$

where $\{\phi_k\}_{k=0}^n$ are built on the choice coefficients of M_n by the recurrence relations in Theorem 4.1.

Proof Obviously,

$$h(\mu, z) = -\log \mid G_\mu(z) \mid$$

and by Theorem 4.14,

$$h(\mu, z) = \log(1- \mid z \mid^2)^{\frac{1}{2}} + \frac{1}{2}\log\sum_{k=0}^{\infty} \mid \phi_k(z) \mid^2 \tag{4.18}$$

where $\{\phi_k\}_{k\geq 0}$ are the orthogonal polynomials of μ. Of course, the measures having the first $n+1$ Fourier coefficients $1, s_1, \cdots, s_n$ have the same choice parameters $g_1, \cdots, g_n$, and as by (4.18)

$$h(\mu, z) \geq \log(1- \mid z \mid^2)^{\frac{1}{2}} + \frac{1}{2}\log \sum_{k=0}^{n} \mid \phi_k(z) \mid^2,$$

we have only to determine a choice sequence $\{g_1, g_2, \cdots, g_n, g_{n+1}(z), \cdots\}$ such that for every fixed $z, \phi_{n+1}(z) = 0$, $\phi_{n+2}(z) = 0, \cdots$.

From the recurrence relations in Theorem 4.1 it is easy to see that the only possibility is $g_{n+1}(z) = \overline{z\phi_n(z)/\phi_n^{\mathrm{T}}(z)}$, $g_{n+k}(z) = 0$ for $k > 1$ and by Corollary 4.5,

$$\{g_1, g_2, \cdots, g_n, \overline{\frac{z\phi_n(z)}{\phi_n^{\mathrm{T}}(z)}}, 0, 0, \cdots\}$$

is indeed a choice sequence for every $\mid z \mid< 1$. ∎

We end this section with another important idea in Szegö's theory.

Proposition 4.17 A positive measure μ on $\mathbf{T}$ has the form

$$\mu(f) = \frac{1}{2\pi}\int_0^{2\pi} \frac{1}{\mid p^T(e^{it}) \mid^2} f(t)dt, \quad f \in C(\mathbf{T}) \tag{4.19}$$

with p a polynomial of degree m and zeros in $\mathbf{D}$, if and only if $\mid g_n \mid< 1$ for every $n \geq 1$ and $g_n = 0$ for $n > m$.

Proof We prove first that if μ has the form (4.19) then its orthogonal polynomials are, for $n \geq m$,

$$\phi_n(z) = z^{n-m}p(z). \tag{4.20}$$

Indeed, for $n \geq m$, $p < n$,

$$\frac{1}{2\pi}\int_0^{2\pi} e^{-ipt}\phi_n(e^{it})d\mu(t) = \frac{1}{2\pi i}\int_{|z|=1} \frac{z^{n-p-1}}{z^m \overline{p(z)}}dz.$$

But $1/z^m\overline{p(z)}$, $\mid z \mid= 1$, has analytic continuation in $\mathbf{D}$, so the last integral is null. Obviously,

$$\frac{1}{2\pi}\int_0^{2\pi} \mid \phi_n(e^{it}) \mid^2 d\mu(t) = 1$$

and the polynomials given by (4.20) are the orthogonal polynomials of μ for $n \geq m$. Now, by Remark 4.9,

$$g_n = -\overline{\Phi_n(0)} = 0 \quad \text{for } n > m.$$

Conversely, suppose $g_n = 0$ for $n > m$. Then by Theorem 4.1

$$\phi_n(z) = z^{n-m}\phi_m(z), \quad n \geq m.$$

Then, $\phi_n^{\mathrm{T}}(z) = \phi_m^{\mathrm{T}}(z)$ for $n \geq m$ and by Theorem 4.14,

$$G_\mu(z) = \frac{1}{\mid \phi_m^{\mathrm{T}}(z) \mid^2}. \tag{4.21}$$

Moreover, by Theorem 1.1, the function F associated to μ by (4.15) is

$$F(z) = \frac{\mathcal{A}_{m-1}(z)}{\mathcal{B}_{m-1}(z)}$$

and the function G associated to μ by (4.14) is then

$$\begin{aligned} G(z) &= \frac{\mathcal{B}_{m-1}(z) + z\mathcal{A}_{m-1}(z)}{\mathcal{B}_{m-1}(z) - z\mathcal{A}_{m-1}(z)} \\ &= \frac{\Psi_m^T(z)}{\Phi_m^T(z)}. \end{aligned}$$

Consequently, $\mid G(z) \mid \leq M$ for $z \in \mathbf{D}$, which shows, by means of (4.14) that μ is absolutely continuous with respect to Lebesgue measure and, by (4.21), it has the required form with $p = \phi_m$. ∎

Note the following consequence of Proposition 4.17.

Corollary 4.18 For $f \in C(\mathbf{T})$

$$\lim_{n\to\infty} \frac{1}{2\pi}\int_0^{2\pi} \frac{f(t)}{\mid \phi_n(e^{it} \mid^2} dt = \frac{1}{2\pi}\int_0^{2\pi} f(t)d\mu(t).$$

Proof Let $\{g_n\}_{n=1}^\infty$ be the choice parameters of μ. Then the measure μ_n associated to the parameters $\{g_1, \cdots, g_n, 0, \cdots\}$ has the form described in Proposition 4.17 with $p = \phi_n$ and the first n Fourier coefficients of μ and μ_n coincide. Consequently,

$$\frac{1}{2\pi}\int_0^{2\pi} \frac{f(t)}{\mid \phi_n(e^{it} \mid^2} dt = \frac{1}{2\pi}\int_0^{2\pi} f(t)d\mu(t), \tag{4.22}$$

for every trigonometric polynomial of degree $\leq n$. Then, use the uniform density of the trigonometric polynomials in $C(\mathbf{T})$. ∎

We present now some extensions of Szegö's theory to the case when μ satisfies the Erdös–Turán condition

$$\mu' > 0 \quad \text{a.e. on } \mathbf{T}. \tag{4.23}$$

A geometric approach is not available in this case, so we start with some estimations. First of all, we have to remove the singular part of μ.

Lemma 4.19 Let γ be a finite Borel measure on $\mathbf{T}$ singular with respect to Lebesgue measure on $\mathbf{T}$. Then there exists a sequence $\{f_n\}_{n=1}^{\infty}$ of functions in $C(\mathbf{T})$ such that $0 < f_n(t) \leq 1$ for all t, $\lim_{n\to\infty} f_n(t) = 1$ a.e. and $\lim_{n\to\infty} \int_0^{2\pi} f_n(t)d\gamma(t) = 0$.

Proof Let $E \subset [0, 2\pi]$ be a Borel set with Lebesgue measure 0 and $\gamma(\mathbf{T} - E) = 0$. Let $\{E_n\}_{n=1}^{\infty}$ be a decreasing sequence of open sets with $E \subset E_n \subset (0, 2\pi)$; the Lebesgue measures of E_n are bounded by 2 and converge to 0.

For $t \in [0, 2\pi]$ define $g_n(t) = \inf\{\mid t - s \mid \mid s \in [0, 2\pi] - E_n\}$. Then g_n are continuous functions on $[0, 2\pi]$, $g_n(t) > 0$ for $t \in E_n$ (E_n are open sets) and $g_n(t) = 0$ for $t \in [0, 2\pi] - E_n$. Moreover, $g_n(t) < 1$ for all t. Define $f_{n,k}(t) = (1 - g_n(t))^k$. Then $\lim_{k\to\infty} f_{n,k}(t) = 0$ for all $t \in E$ and $0 \leq f_{n,k} \leq 1$. By Lebesgue's dominated convergence theorem, there exists an integer k_n with

$$0 \leq \int_0^{2\pi} f_{n,k_n}(t)d\gamma(t) \leq \frac{1}{n}$$

so that, defining $f_n = f_{n,k_n}$ we obtain functions with the required properties. Indeed, the first one and the third are obviously satisfied. As $f_n(t) = 1$ for $t \notin E_n$, $\lim_{n\to\infty} f_n(t) = 1$ for $t \in \bigcap_{n=1}^{\infty} E_n$, but the Lebesgue measure of $\bigcap_{n=1}^{\infty} E_n$ is zero. Finally, as $E_n \subset (0, 2\pi)$, $f_n(0) = f_n(2\pi) = 1$, so $f_n \in C(\mathbf{T})$. ∎

Thus an estimation of the Szegö's paramaters is obtained.

Lemma 4.20 There exists an absolute constant C such that

$$\mid \Phi_{n+1}(0) \mid \leq C \int_0^{2\pi} \mid \frac{\mid \phi_n(e^{it}) \mid^2}{\mid \phi_{n+1}(e^{it}) \mid^2} - 1 \mid dt.$$

Proof By the recurrence formulas in Theorem 4.1,

$$| \Phi_{n+1}(0) |=| \frac{\Phi_{n+1}^{\mathrm{T}}(e^{it})}{\Phi_n^{\mathrm{T}}(e^{it})} - 1 |$$

and as $| \Phi_{n+1}(0) |< 1$, we get $| \Phi_{n+1}^{\mathrm{T}}(e^{it}) |\leq 2 | \Phi_n^{\mathrm{T}}(e^{it}) |$. Consequently

$$| \Phi_{n+1}(0) |\leq 2 | \frac{\Phi_n^{\mathrm{T}}(e^{it})}{\Phi_{n+1}^{\mathrm{T}}(e^{it})} - 1 | . \tag{4.24}$$

Also by the recurrence relations,

$$\frac{\Phi_n^{\mathrm{T}}(z)}{\Phi_{n+1}^{\mathrm{T}}(z)} - 1 = -z\overline{\Phi_{n+1}(0)}\frac{\Phi_n(z)}{\Phi_{n+1}^{T}(z)}, \quad z \in \bar{\mathbf{D}}.$$

Consequently

$$| \frac{\Phi_n^{\mathrm{T}}(e^{it})}{\Phi_{n+1}^{\mathrm{T}}(e^{it})} |^2 -2\,\mathrm{Re}(\frac{\Phi_n^{\mathrm{T}}(e^{it})}{\Phi_{n+1}^{\mathrm{T}}(e^{it})}) + 1 = (1 - \frac{k_n^2}{k_{n+1}^2}) | \frac{\Phi_n^{\mathrm{T}}(e^{it})}{\Phi_{n+1}^{\mathrm{T}}(e^{it})} |^2$$

where we have used (4.9). Then

$$\mathrm{Re}(\frac{\Phi_n^{\mathrm{T}}(e^{it})}{\Phi_{n+1}^{\mathrm{T}}(e^{it})} - 1) = \frac{1}{2}(\frac{| \phi_n(e^{it}) |^2}{| \phi_{n+1}(e^{it}) |^2} - 1).$$

Now, observe that $\Phi_n^{\mathrm{T}}/\Phi_{n+1}^{\mathrm{T}} - 1$ is analytic in $\bar{\mathbf{D}}$ and vanishes at $z = 0$. Denote by u the harmonic function $\mathrm{Re}(\Phi_n^{\mathrm{T}}/\Phi_{n+1}^{\mathrm{T}} - 1)$ and use Kolmogorov's inequality in order to obtain (here, and only here, ~ denotes the complex conjugate function):

$$\begin{aligned}
&(\int_0^{2\pi} | \frac{\Phi_n^{\mathrm{T}}(e^{it})}{\Phi_{n+1}^{\mathrm{T}}(e^{it})} - 1 |^{\frac{1}{2}}\, dt)^2 = (\int_0^{2\pi} | u + i\tilde{u} |^{\frac{1}{2}}\, dt)^2 \\
&\leq (\int_0^{2\pi} (| u | + | \tilde{u} |)^{\frac{1}{2}} dt)^2 \leq (\int_0^{2\pi} (| u |^{\frac{1}{2}} + | \tilde{u} |^{\frac{1}{2}})dt)^2 \\
&\leq 2((\int_0^{2\pi} | u |^{\frac{1}{2}}\, dt)^2 + (\int_0^{2\pi} | \tilde{u} |^{\frac{1}{2}}\, dt)^2 \leq 2(2\pi + B) \int_0^{2\pi} | u |\, dt \\
&= (2\pi + B) \int_0^{2\pi} | \frac{| \phi_n(e^{it}) |^2}{| \phi_{n+1}(e^{it}) |^2} - 1 |\, dt
\end{aligned}$$

and by (4.24) the required inequality follows. ∎

The following result plays a key technical role.

Proposition 4.21 If $\mu' > 0$ a.e., then

$$\lim_{n\to\infty} \int_0^{2\pi} \mid \frac{\mid \phi_n(e^{it}) \mid^2}{\mid \phi_{n+1}(e^{it}) \mid^2} - 1 \mid dt = 0.$$

Proof Let $f \in C(\mathbf{T})$, $f \geq 0$. By Hölder's inequality,

$$\begin{aligned}
&\frac{1}{2\pi}\int_0^{2\pi} (\mu'(t)f(t))^{\frac{1}{4}} dt \\
&= \frac{1}{2\pi}\int_0^{2\pi} \mid \frac{\phi_n(e^{it})}{\phi_{n+1}(e^{it})} \mid^2 (\mid \phi_{n+1}(e^{it}) \mid^2 \mu'(t))^{\frac{1}{4}} \frac{(f(t))^{\frac{1}{4}}}{\mid \phi_n(e^{it}) \mid^{\frac{1}{2}}} dt \\
&\leq (\frac{1}{2\pi}\int_0^{2\pi} \frac{\mid \phi_n(e^{it}) \mid}{\mid \phi_{n+1}(e^{it}) \mid} dt)^{\frac{1}{2}} (\frac{1}{2\pi}\int_0^{2\pi} \mid \phi_{n+1}(e^{it}) \mid^2 \mu'(t) dt)^{\frac{1}{4}} (\frac{1}{2\pi}\int_0^{2\pi} \frac{f(t)}{\mid \phi_n(e^{it}) \mid^2} dt)^{\frac{1}{4}}.
\end{aligned}$$

The second factor on the right–hand side is less than 1 and by Corollary 4.18, we obtain

$$\begin{aligned}
(\frac{1}{2\pi}\int_0^{2\pi} (\mu'(t)f(t))^{\frac{1}{4}} dt)^4 \leq (\liminf \frac{1}{2\pi}\int_0^{2\pi} \frac{\mid \phi_n(e^{it}) \mid}{\mid \phi_{n+1}(e^{it}) \mid} dt)^2 \\
\cdot (\frac{1}{2\pi}\int_0^{2\pi} f(t) d\mu(t)). \qquad (4.25)
\end{aligned}$$

We can choose a family $\{f_m\}_{m\geq 1} \subset C(\mathbf{T})$ as in Lemma 4.19 with respect to the singular part of μ. Moreover, for fixed $\epsilon > 0$ we consider a family $\{g_k\}_{k\geq 1} \subset C(\mathbf{T})$ such that $0 < g_k(t) \leq 1/\epsilon$ and $\lim\limits_{k\to\infty} g_k(t) = (\mu'(t)+\epsilon)^{-1}$ a.e. Here is the point where we use the Erdös–Turán condition. Since $\lim\limits_{\epsilon\to\infty} \dfrac{\mu'(t)}{\mu'(t)+\epsilon} = 1$ for each t with $\mu'(t) > 0$, in our case this happens almost everywhere. We have

$$\begin{aligned}
&\frac{1}{2\pi}\int_0^{2\pi} f_m(t)g_k(t)d\mu(t) \overset{m\to\infty}{\to} \frac{1}{2\pi}\int_0^{2\pi} g_k(t)\mu'(t)dt \\
&\overset{k\to\infty}{\to} \frac{1}{2\pi}\int_0^{2\pi} \frac{\mu'(t)}{\mu'(t)+\epsilon} dt \overset{\epsilon\to 0}{\to} 1
\end{aligned}$$

and

$$\begin{aligned}
&\frac{1}{2\pi}\int_0^{2\pi} (\mu'(t)f_m(t)g_k(t))^{\frac{1}{4}} dt \overset{m\to\infty}{\to} \frac{1}{2\pi}\int_0^{2\pi} (\mu'(t)g_k(t))^{\frac{1}{4}} dt \\
&\overset{k\to\infty}{\to} \frac{1}{2\pi}\int_0^{2\pi} (\frac{\mu'(t)}{\mu'(t)+\epsilon})^{\frac{1}{4}} dt \overset{\epsilon\to 0}{\to} 1.
\end{aligned}$$

Therefore, by (4.25)

$$\liminf \frac{1}{2\pi}\int_0^{2\pi} \frac{|\phi_n(e^{it})|^2}{|\phi_{n+1}(e^{it})|^2} dt \geq 1.$$

Using (4.22)

$$\frac{1}{2\pi}\int_0^{2\pi} \frac{|\phi_n(e^{it})|^2}{|\phi_{n+1}(e^{it})|^2} dt = \frac{1}{2\pi}\int_0^{2\pi} |\phi_n(e^{it})|^2 \, d\mu(t) = 1$$

so that

$$\lim_{n\to\infty} \frac{1}{2\pi}\int_0^{2\pi} \left(\frac{|\phi_n(e^{it})|}{|\phi_{n+1}(e^{it})|} - 1\right)^2 dt = 0. \tag{4.26}$$

By Schwartz's inequality, we get

$$\left(\frac{1}{2\pi}\int_0^{2\pi} \left| \frac{|\phi_n(e^{it})|^2}{|\phi_{n+1}(e^{it})|^2} - 1 \right| dt\right)^2$$
$$\leq \frac{1}{2\pi}\int_0^{2\pi} \left(\frac{|\phi_n(e^{it})|}{|\phi_{n+1}(e^{it})|} + 1\right)^2 dt \cdot \frac{1}{2\pi}\int_0^{2\pi} \left(\frac{|\phi_n(e^{it})|}{|\phi_{n+1}(e^{it})|} - 1\right)^2 dt$$

and the result follows from (4.22) and (4.26). ∎

In the context of Szegö's theory, the condition (4.16) immediately implies that $\Phi_n(0) \to 0$. It is a remarkable result that this holds also in the presence of condition (4.23).

Theorem 4.22 If $\mu' > 0$ a.d., then

$$\lim_{n\to\infty} \Phi_n(0) = 0.$$

Proof Clearly we have only to use Lemma 4.20 and Proposition 4.21. ∎

In view of the relation between μ' and $G\mu$ ($|G\mu(e^{it})|^2 = \mu'(t)$, of Corollary 1.3) in the case when Szegö's condition holds, the following result provides a natural extension of Theorem 4.14 (1).

Theorem 4.23 If $\mu' > 0$ a.e., then

$$\lim \frac{1}{2\pi}\int_0^{2\pi} (|\phi_n(e^{it})| (\mu'(t))^{\frac{1}{2}} - 1)^2 dt = 0$$

Proof We have that

$$\begin{aligned}
0 &\leq \frac{1}{2\pi}\int_0^{2\pi}(\mid \phi_n(e^{it}) \mid (\mu(t))^{\frac{1}{2}} - 1)^2 dt \\
&= \frac{1}{2\pi}\int_0^{2\pi} \mid \phi_n(e^{it}) \mid^2 \mu'(t)dt - \frac{1}{\pi}\int_0^{2\pi} \mid \phi_n(e^{it}) \mid (\mu'(t))^2 dt + 1 \\
&\leq \frac{1}{2\pi}\int_0^{2\pi} \mid \phi_n(e^{it}) \mid^2 d\mu(t) - \frac{1}{\pi}\int_0^{2\pi} \mid \phi_n(e^{it}) \mid (\mu'(t))^{\frac{1}{2}} dt + 1 \\
&= 2 - \frac{1}{2\pi}\int_0^{2\pi} \mid \phi_n(e^{it}) \mid (\mu^2(t))^{\frac{1}{2}} dt.
\end{aligned}$$

Consequently, it will be sufficient to prove that

$$\liminf \frac{1}{2\pi}\int_0^{2\pi} \mid \phi_n(e^{it}) \mid (\mu'(t))^{\frac{1}{2}} dt \geq 1.$$

For this purpose, let $f \in c(\mathbf{T})$ be positive. By Hölder's inequality,

$$\begin{aligned}
&(\frac{1}{2\pi}\int_0^{2\pi}(\mu'(t)f(t))^{\frac{1}{4}} dt)^4 \\
&\leq (\frac{1}{2\pi}\int_0^{2\pi} \mid \phi_n(e^{it}) \mid (\mu'(t))^{\frac{1}{2}} dt)^2 (\frac{1}{2\pi}\int_0^{2\pi} \frac{f(t)}{\mid \phi_n(e^{it}) \mid^2} dt).
\end{aligned}$$

Letting $n \to \infty$ and using Corollary 4.18, we get

$$\begin{aligned}
&(\frac{1}{2\pi}\int_0^{2\pi}(f(t)\mu'(t))^{\frac{1}{4}} dt)^4 \\
&\leq \liminf(\frac{1}{2\pi}\int_0^{2\pi} \mid \phi_n(e^{it}) \mid (\mu'(t))^{\frac{1}{2}} dt)^2 (\frac{1}{2\pi}\int_0^{2\pi} f(t)d\mu(t)).
\end{aligned}$$

Now we can finish the proof exactly as in Proposition 4.21 by taking the sequences $\{f_m\}_{m\geq 1}$ and $\{g_k\}_{k\geq 1}$ as there. ■

4.4 Szegö's limit theorems

Szegö's two limit theorems concern the asymptotic properties of the positive Toepelitz determinants, having interpretation in the geometry of the stationary process represented by the considered Toeplitz form.

Let μ be a positive measure on $\mathbf{T}$, $G\mu$ its spectral factor and $\{g_n\}_{n\geq 1}$ the parameters of μ. We suppose that $\mu(1) = 1$ and denote by M_n the Toeplitz matrices based on the

Fourier coefficients of μ. Moreover suppose that $\det M_n \neq 0$ for $n \geq 1$. The first Limit Theorem of Szegö is the following.

Theorem 4.24

$$\lim_{n\to\infty} \frac{\det M_{n+1}}{\det M_n} = G_\mu^2(0) = \prod_{n=1}^{\infty}(1-\mid g_n\mid^2) = \exp(\frac{1}{2\pi}\int_0^{2\pi}\log\mu'(t)dt).$$

Proof By formula (2.27),

$$\det M_n = \prod_{k=1}^{n}(1-\mid g_k\mid^2)^{n-k}.$$

Consequently,

$$\frac{\det M_{n+1}}{\det M_n} = \prod_{k=1}^{n+1}(1-\mid g_k\mid^2)$$

and now we have to use (3.1) and Theorem 3.15. ∎

This result has a matricial counterpart. Take μ to be a semispectral measure ($\mu(1) = I$) with values in $\mathcal{L}(\mathcal{H})$ and let M_n be the Toeplitz block matrices based on its Fourier coefficients. $\{G_n\}_{n=1}^{\infty}$ are the parameters of μ associated by Theorem 2.4 and $G\mu$ is its spectral factor. Suppose $\dim\mathcal{H} < \infty$. In this case we obtain the following variant of Theorem 4.24.

Theorem 4.25

$$\lim_{n\to\infty} \frac{\det M_{n+1}}{\det M_n} = \prod_{n=1}^{\infty}\det D_{G_n}^2 = \det G_\mu^2(0).$$

Proof Again by formula (2.27),

$$\det M_n = \prod_{k=1}^{n}\det D_{G_k}^{2(n-k)}$$

and then the expression of the limit in terms of the parameters $\{G_n\}$ is obvious. For the expression in terms of the spectral factor we have to use Theorem 3.15. ∎

Remark 4.26 In view of Theorem 4.15, the first Szegö Theorem has a geometric interpretation. Thus,

$$\inf_{\substack{p\in\mathcal{P}_{n+1}\\ p(0)=1}} \frac{1}{2\pi}\int_0^{2\pi} | p(e^{it}) |^2 \, d\mu(t) = \frac{1}{\sum_{k=0}^{n+1} | \phi_k(0) |^2}$$

$$= \prod_{k=1}^{n+1}(1- | g_n |^2) = \frac{\det M_{n+1}}{\det M_n}$$

so that Theorem 4.24 is another way of reflecting the convergence of the angles between $\tilde{\mathcal{K}}_n$ and the present to the angle between past and present (according to the terminology in Theorem 4.15). ∎

The second Szegö Limit Theorem (the Strong Szegö Limit Theorem) describes the asymptotic behaviour of the determinants M_n themselves. First, we establish a formula for this limit in terms of the coefficients $\{G_n\}$ of the semispectral measure μ with values in $\mathcal{L}(\mathcal{H})$, $\dim\mathcal{H} < \infty$.

Theorem 4.27

$$\lim_{n\to\infty} \frac{\det M_n}{G_\mu^{2n+2}(0)} = \exp\left(\sum_{n=1}^{\infty} n \log \det D_{G_n}^{-2}\right).$$

Proof By formula (2.27) and Theorem 3.15,

$$\lim_{n\to\infty} \frac{\det M_n}{G_\mu^{2n+2}(0)}$$

$$= \frac{1}{\prod_{n=1}^{\infty} \det D_{G_n}^{2n}} = \exp\left(\sum_{n=1}^{\infty} n \log \det D_{G_n}^{-2}\right)$$

with the observation that the limit (which always exists) is finite if and only if the infinite product $\prod_{n=1}^{\infty} \det D_{G_n}^{2n}$ converges. ∎

In the classical form of Theorem 4.27 one computes the limit with the aid of the spectral factor. The core of the result is the following.

Proposition 4.28 Let μ be a positive measure on $\mathbf{T}$ of the form (4.19) with p a polynomial of degree m and zeros in $\mathbf{D}$. Then, for $n \geq m$,

$$\frac{\det M_{m-1}}{G_\mu^{2m}(0)} = \frac{\det M_n}{G_\mu^{2n+2}(0)} = \exp(\frac{1}{\pi} \iint_{|z|\leq 1} | G'_\mu(z)/G_\mu(z) |^2 \, d\sigma.$$

Proof We begin by remarking that we can choose $G_\mu = 1/p^{\mathrm{T}}$. By Proposition 4.17, formula (2.27) and Theorem 3.15 we have

$$\frac{\det M_n}{G_\mu^{2n+2}(0)} = \frac{\det M_{m-1}}{G_\mu^{2m}(0)}$$

for $n \geq m$. The rest of the proof is devoted to the computation of the two remaining terms with the aid of the zeros of p^{T}. Consider $q(z) = p^{\mathrm{T}}(z) = c(z - z_1)\cdots(z - z_m)$, with $| z_k |> 1$, $k = 1, \cdots, m$.

First, we compute $\exp(\frac{1}{\pi} \iint_{|z|\leq 1} | G'_\mu(z)/G_\mu(z) |^2 \, d\sigma)$. We have

$$\frac{G'_\mu(z)}{G\mu(z)} = -\frac{q'(z)}{q(z)} = -(\frac{1}{z - z_1} + \cdots + \frac{1}{z - z_m})$$

so that

$$\frac{1}{\pi} \iint_{|z|\leq 1} | \frac{G'_\mu(z)}{G_\mu(z)} |^2 \, d\sigma = \sum_{j,k=1}^{m} \frac{1}{\pi} \iint_{|z|\leq 1} \frac{d\sigma}{(z - z_j)(\bar{z} - \bar{z}_k)}.$$

Passing to polar coordinates and using Cauchy's formula,

$$\begin{aligned}
\iint_{|z|\leq 1} \frac{d\sigma}{(z - z_j)(\bar{z} - \bar{z}_k)} &= \int_0^1 \int_0^{2\pi} \frac{r\,dr\,dt}{(re^{it} - z_j)(re^{it} - \bar{z}_k)} \\
&= \int_0^1 r \int_0^{2\pi} \frac{dt}{(re^{it} - z_j)(re^{-it} - \bar{z}_k)} dr \\
&= 2\pi \int_0^1 \frac{r\,dr}{z_j \bar{z}_k - r^2} \\
&= \pi \log \frac{z_j \bar{z}_k}{z_j \bar{z}_k - 1}
\end{aligned}$$

and, finally,

$$\exp(\frac{1}{\pi} \iint_{|z|\leq 1} | \frac{G'_\mu(z)}{G_\mu(z)} |^2 \, d\sigma)$$

$$= \prod_{j,k=1}^{m} \frac{z_j \bar{z}_k}{z_j \bar{z}_k - 1}$$

$$= \frac{| z_1 \cdots z_m |^{2m}}{\prod\limits_{j,k=1}^{m} (z_j \bar{z}_k - 1)}.$$

The computation of $\dfrac{\det M_m}{G^{2m+2}(0)}$ is based on a change of coordinates using Lagrange interpolation. We have for $u = (u_1, \cdots, u_m) \in \mathbf{C}^m$,

$$(M_{m-1}u, u) = \frac{1}{2\pi} \int_0^{2\pi} | \sum_{k=1}^{m} u_k e^{-i(k-1)t} |^2 \frac{1}{| q(e^{it}) |^2} dt$$

$$= \frac{1}{2\pi} \int_0^{2\pi} | \frac{u(e^{it})}{q(e^{it})} |^2 \, dt$$

with $u(z) = u_1 z^{m-1} + u_2 z^{m-2} + \cdots + u_{m-1} z + u_m$. Now, using Lagrange's formula

$$\frac{u(z)}{q(z)} = \sum_{k=1}^{m} \frac{u(z_k)}{q'(z_k)} \frac{1}{z - z_k}.$$

Consider

$$v_k = \frac{u(z_k)}{q'(z_k)} = \frac{1}{q'(z_k)} \sum_{s=1}^{m} u_s z_k^{m-s}, \quad k = 1, \cdots, m.$$

The required change of coordinates is exactly

$$\{u_k\}_{k=1}^{m} \;\rightarrow\; \{v_k\}_{k=1}^{m}$$

whose Jacobian is

$$J = \frac{\partial(v_1, \cdots, v_m)}{\partial(u_1, \cdots, u_m)} = \frac{\prod\limits_{kj} (z_k - z_j)}{q'(z_1) \cdots q'(z_m)}$$

$$= (-1)^{\frac{1}{2}m(m+1)} c^{-m} \prod_{kj} (z_k - z_j)^{-1}.$$

In this way,

$$(M_{m-1}u, u) = \sum_{j,k=1}^{m} v_j \bar{v}_k \frac{1}{2\pi} \int_0^{2\pi} \frac{dt}{(e^{it} - z_j)(e^{it} - \bar{z}_k)}$$

$$= \sum_{j,k=1}^{m} v_j \bar{v}_k \frac{1}{z_j \bar{z}_k - 1}$$

by Cauchy's formula. Now we can compute

$$\det M_{m-1} = \mid J \mid^2 \det((\frac{1}{z_j \bar{z}_k - 1})_{j,k=1}^{m})$$

$$= \mid c \mid^{-2m} \prod_{k<j} \mid z_k - z_j \mid^{-2} \frac{\prod_{k<j} \mid z_k - z_j \mid^2}{\prod_{j,k=1}^{m} (z_j \bar{z}_k - 1)}$$

$$= \mid c \mid^{-2m} \frac{1}{\prod_{j,k=1}^{m} (z_j z_k - 1)}$$

and we obtain

$$\frac{\det M_{m-1}}{G_\mu^{2m}(0)} = \frac{\mid z_1 \cdots z_m \mid^2}{\prod_{j,k=1}^{m} (z_j \bar{z}_k - 1)} = \exp(\frac{1}{\pi} \int\int_{|z|\leq 1} \mid \frac{G'_\mu(z)}{G_\mu(z)} \mid^2 d\sigma). \qquad \blacksquare$$

We can now describe one of the many variants (regarding the finiteness conditions) of the strong Szegö Limit Theorem with the limit expressed in terms of the spectral factor. Take $a \in L^1$ to be a positive function and consider M_n the positive Toeplitz matrices based on its Fourier coefficients. Let G be the spectral factor of the measure having a as density and $\{\phi_n\}_{n\geq 0}$ be the orthogonal polynomials of this measure.

Theorem 4.29 If

$$\sum_{n=1}^{\infty} n \mid \phi_n(0) \mid^2 < \infty$$

then the following limit exists and is finite:

$$\lim_{n\to\infty} \frac{\det M_n}{G^{2n+2}(0)} = \exp(\frac{1}{\pi} \int\int_{|z|\leq 1} \mid \frac{G'(z)}{G(z)} \mid^2 d\sigma).$$

Proof Consider the measure

$$\mu(f) = \frac{1}{2\pi} \int_0^{2\pi} a(t) f(t) dt, \quad f \in C(\mathbf{T}).$$

Then $\{\phi_n\}_{n\geq 0}$ are its orthogonal polynomials and $\{g_n\}_{n\geq 1}$ its choice parameters. Consider the measure μ_n with the parameters $\{g_1, \cdots, g_n, 0, \cdots\}$; then, according to Proposition 4.17, μ_n has the form (4.19) with $p = \phi_n$. The main remark is the following:

$$\log \frac{\det M_{m-1}(\mu_n)}{G^{2n}_{\mu_n}(0)} = \log \frac{\det M_n(\mu_n)}{G^{2n}_{\mu_n}(0)}$$

$$= \sum_{k=1}^{n} \log \frac{\det M_k(\mu_n)}{\det M_{k-1}(\mu_n)} + \log\det M_o(\mu_n) - (n+1)\log G^2_{\mu_n}(0)$$

$$= \sum_{k=0}^{n} \log \frac{\sum\limits_{p=0}^{n} \mid \phi_p(0) \mid^2}{\sum\limits_{p=0}^{k} \mid \phi_p(0) \mid^2}$$

$$= -\sum_{k=0}^{n-1} \log(1 - \frac{\sum\limits_{p=k+1}^{n} \mid \phi_p(0) \mid^2}{\sum\limits_{p=0}^{n} \mid \phi_p(0) \mid^2})$$

for $n \geq m$. Denote

$$d = \frac{\sum\limits_{p=k+1}^{n} \mid \phi_p(0) \mid^2}{\sum\limits_{p=0}^{n} \mid \phi_p(0) \mid^2}.$$

Then

$$-\log(1 - \frac{\sum\limits_{p=k+1}^{n} \mid \phi_p(0) \mid^2}{\sum\limits_{p=0}^{n} \mid \phi_p(0) \mid^2})$$

$$= \sum_{p=1}^{\infty} \frac{d^p}{p} \leq \frac{d + d^2}{1-d}$$

$$\leq \frac{2d}{1-d} \leq C_1 d.$$

Consequently, for $n \geq m$,

$$\log \frac{\det M_n(\mu_n)}{G^{2m+2}_{\mu_n}(0)}$$

$$\leq \frac{C_1}{\sum\limits_{p=0}^{n} \mid \phi_p(0) \mid^2} \sum_{k=0}^{n-1} \sum_{p=k+1}^{n} \mid \phi_p(0) \mid^2$$

$$\leq \frac{C_1}{\mid \phi_o(0) \mid^2} \sum_{k=1}^{\infty} k \mid \phi_k(0) \mid^2 = C_2 \sum_{k=1}^{\infty} k \mid \phi_k(0) \mid^2 .$$

Now, we already know that $\{\frac{\det M_n}{G^{2n+2}(0)}\}_{n\geq 1}$ is an increasing sequence, and

$$\lim_{n\to\infty}\frac{\det M_n}{G^{2n+2}(0)} = \lim_{n\to\infty}\frac{\det M_n(\mu_n)}{G^{2n+1}_{\mu_n}(0)}$$

$$=\lim_{n\to\infty}\exp\left(\frac{1}{\pi}\int\int_{midz|\leq 1} \left| \frac{(\phi_n^{\mathrm{T}})'(z)}{\phi_n^{\mathrm{T}}(z)} \right|^2 d\sigma\right)$$

by Proposition 4.28. But, on the one hand,

$$\int\int_{|z|\leq 1} \left| \frac{(\phi_n^{\mathrm{T}})'(z)}{\phi_n^{\mathrm{T}}(z)} \right|^2 d\sigma \leq C_3 \sum_{k=1}^{\infty} k \mid \phi_k(0) \mid^2$$

and on the other hand, by Theorem 4.14, ϕ_n^{T} converge uniformly on compact subsets of $\mathbf{D}$ to G^{-1}; therefore $(\phi_n^{\mathrm{T}})'$ converge uniformly on compact subsets of $\mathbf{D}$ to $(G^{-1})'$. By Fatou's lemma we conclude that

$$\lim_{n\to\infty}\frac{\det M_n}{G^{2n+2}(0)} = \exp\left(\frac{1}{\pi}\int\int_{|z|\leq 1} \left| \frac{G'(z)}{G(z)} \right|^2 d\sigma\right)$$

and the right–hand side is finite. ∎

4.5 Operator–valued orthogonal polynomials

In this section we are concerned with some algebraic and asymptotic properties of operator–valued orthogonal polynomials. We intensively use the choice parameters of the semispectral measure we started with and in this way most of the considerations become transparent adaptations of the scalar case, so that we will only illustrate some representative aspects.

Let $\mathcal{H}$ be a separable Hilbert space and μ an $\mathcal{L}(\mathcal{H})$–valued semispectral measure on $\mathbf{T}$ with $\mu(1) = I$. Let $C(\mathbf{T}, \mathcal{L}(\mathcal{H}))$ be the set of continuous functions on $\mathbf{T}$ with values in $\mathcal{L}(\mathcal{H})$. We define

$$(f,g)_{1\mu} = \frac{1}{2\pi}\int_0^{2\pi} f(e^{it})d\mu(t)g(e^{it})^* \tag{4.27}$$

$$(f,g)_{r\mu} = \frac{1}{2\pi}\int_0^{2\pi} f(e^{it})^* d\mu(t)g(e^{it}) \tag{4.28}$$

for $f, g \in C(\mathbf{T}, \mathcal{L}(\mathcal{H}))$. Thus, $C(\mathbf{T}, \mathcal{L}(\mathcal{H}))$ becomes a left (right) prehilbertian $\mathcal{L}(\mathcal{H})$–premodule. To associate orthogonal polynomials to μ means finding a sequence of polynomials $\{\phi_n\}_{n\geq 0}$ having the properties:

$$\phi_n(z) = L_{nn}z^n + L_{n,n-1}z^{n-1} + \cdots + L_{n0}, \quad L_{nk} \in \mathcal{L}(\mathcal{H}), \quad L_{nn} \geq 0, \tag{4.29}$$

$$(\phi_n, \phi_m)_{1\mu} = \delta_{nm} I \tag{4.30}$$

(these are the left orthogonal polynomials of μ) and, unlike the scalar case, a sequence of polynomials $\{\hat{\phi}_n\}_{n\geq 0}$ having the properties:

$$\hat{\phi}_n(z) = R_{nn}z^n + R_{n,n-1}z^{n-1} + \cdots + R_{no}, \quad R_{nk} \in \mathcal{L}(\mathcal{H}), \quad R_{nn} \geq 0, \tag{4.31}$$

$$(\hat{\phi}_n, \hat{\phi}_m)_{r\mu} = \delta_{nm} I \tag{4.32}$$

(these are the right orthogonal polynomials of μ).

Denoting by $S_n = \mu(\chi_{-n})$, $n \in \mathbf{Z}$, the Fourier coefficients of μ, we remark that as in the scalar case, the existence of the orthogonal polynomials of μ is related to the invertibility of the matrices:

$$M_n = \begin{bmatrix} I & S_1 & \cdots & S_n \\ S_1^* & I & \cdots & S_{n-1} \\ \cdot & & & \cdot \\ \cdot & & & \cdot \\ \cdot & & & \cdot \\ S_n^* & S_{n-1}^* & \cdots & I \end{bmatrix}, \quad n \geq 1.$$

A convenient sufficient condition (which in the case $\dim \mathcal{H} < \infty$ is also necessary), is that D_{G_n} are invertible operators for all $n \geq 1$, where $\{G_n\}_{n\geq 1}$ are the choice parameters of μ. We have a first result.

Proposition 4.30 If $\mathring{L} = (\mathring{L}_{no}, \mathring{L}_{n1}, \cdots \mathring{L}_{nk})$ is the solution of the system

$$\mathring{L} M_m = (0_n, I)$$

then

$$L_{nk} = L_{nn}^{-1} \mathring{L}_{nk}. \quad 0 \leq k < n$$

and

$$L_{nn} = (D_{G_1}^{-1} \cdots D_{G_n}^{-2} \cdots D_{G_1}^{-1})^{\frac{1}{2}}.$$

Proof We get

$$\mathring{L} = (0_n, I) M_n^{-1} = (0_n, I) E_n^{-1} E_n^{*-1}$$

where E_n are the operators defined by (2.19) and then by Proposition A.3,

$$E_n^{-1} = \begin{bmatrix} E_{n-1}^{-1} & -E_{n-1}^{-1}U(\{G_k\}_{k=1}^{n-1})\tilde{L}(\{G_k\}_{k=1}^{n})D_{G_1}^{-1}\cdots D_{G_n}^{-1} \\ 0 & D_{G_1}^{-1}\cdots D_{G_n}^{-1} \end{bmatrix}.$$

Consequently,

$$\mathring{L} = (-D_{G_1}^{-1}\cdots D_{G_n}^{-2}\cdots D_{G_1}^{-1}\tilde{L}(\{G_k\}_{k=1}^{n})^{*}U(\{G_k\}_{k=1}^{n-1})^{*}(E_{n-1}^{-1})^{*}, D_{G_1}^{-1}\cdots D_{G_n}^{-1}\cdots D_{G_1}^{-1}). \tag{4.33}$$

Thus, $\mathring{L}_{nn} = D_{G_1}^{-1}\cdots D_{G_n}^{-1}\cdots D_{G_1}^{-1}$ and we immediately verify that

$$L_{nk} = (D_{G_1}\cdots D_{G_n}^{2}\cdots D_{G_1})^{\frac{1}{2}}\mathring{L}_{nk}, \quad 0 \leq k \leq n$$

are the solutions of the system

$$(L_{no}, \cdots, L_{nn})M_n(I_n \bigoplus L_{nn}) = (0_n, I)$$

which is an equivalent form of the orthogonality relations (4.30). ∎

A similar result holds for the right orthogonal polynomials. Furthermore, define the polynomials

$$\Phi_n(z) = (D_{G_1}\cdots D_{G_n}^{2}\cdots D_{G_1})^{\frac{1}{2}}\phi_n(z)$$
$$\hat{\Phi}_n(z) = \hat{\phi}_n(z)(D_{G_1^*}\cdots D_{G_n^*}^{2}\cdots D_{G_1^*})^{\frac{1}{2}}.$$

Theorem 4.31 The polynomials Φ_n and $\hat{\Phi}_n$ verify the recurrence formulas:

$$\Phi_{n+1}(z) = z\hat{\Phi}_n(z) - D_{G_1}\cdots D_{G_n}G_{n+1}^{*}D_{G_n^*}^{-1}\cdots D_{G_1^*}^{-1}\hat{\Phi}_n^{T}(z)$$
$$\hat{\Phi}_{n+1}(z) = z\hat{\Phi}_n(z) - \Phi_n^{T}(z)D_{G_1}^{-1}\cdots D_{G_n}^{-1}G_{n+1}^{*}D_{G_n^*}\cdots D_{G_1^*}.$$

Proof First, we prove that

$$(\frac{\Phi_{n+1}^{\mathrm{T}}(z) - \Phi_n^{\mathrm{T}}(z)}{z}, \chi_{k})_{r\mu} = 0, \quad 0 \leq k \leq n-1. \tag{4.34}$$

If we denote

$$\Phi_n(z) = z^n + C_{n,n-1}z^{n-1} + \cdots + C_{no}$$

then (4.34) is equivalent to

$$(C_{n+1,n} - C_{n,n-1}, \cdots, C_{n+1,1} - C_{n+1,o})C_{n+1,o})M_n(I_n, 0) = 0_{n+1}. \tag{4.35}$$

Since $(\chi_k, \Phi_{n+1})_{1\mu} = 0$ for $0 \leq k \leq n$, we have

$$(I_{n+1}, 0)M_{n+1}(C^*_{n+1,o}, \cdots, C^*_{n+1,n}, I)^{\mathrm{t}} = 0_{n+2}.$$

Consequently,

$$(C_{n+1,n}, \cdots, C_{n+1,o})M_n = 0_{n+1}. \tag{4.36}$$

Since $(\chi_k, \Phi_n)_{1\mu} = 0$ for $0 \leq k < n$, we have

$$(I_n, 0)M_n(C^*_{no}, \cdots, C^*_{n,n-1}, I)^{\mathrm{t}} = 0_{n+1}.$$

Consequently,

$$(C_{n,n-1}, \cdots, C_{no}, 0)M_n(I_n, 0) = 0_{n+1}. \tag{4.37}$$

From (4.36) and (4.37) we obtain (4.35) and hence (4.34). It follows that

$$\frac{\Phi^{\mathrm{T}}_{n+1}(z) - \Phi^{\mathrm{T}}_n(z)}{z} = \hat{\Phi}_n(z)A$$

with $A \in \mathcal{L}(\mathcal{H})$, and so

$$\Phi_{n+1}(z) = z\Phi_n(z) + A^*\hat{\Phi}^T_n(z).$$

It remains to prove that

$$C_{n+1,o} = -D_{G_1} \cdots D_{G_n} G^*_{n+1} D^{-1}_{G^*_n} \cdots D^{-1}_{G^*_1}. \tag{4.38}$$

For this purpose, we remark that by (4.33) we have to determine the upper element in $E^{-1}_{n-1}U(\{G_k\}^{n-1}_{k=1})\tilde{L}(\{G_k\}^n_{k=1})$. Taking into account the definitions, we immediately find that this element is $D_{G^*_1} \cdots D_{G^*_{n-1}} - L(\{G_k\}^{n-1}_{k=1})D(\{G_k\}^{n-1}_{k=1})^{-1}\tilde{L}(\{G_k\}^{n-1}_{k=1})$, so that it remains to prove the equality:

$$D_{G^*_1} \cdots D_{G^*_n} - L(\{G_k\}^n_{k=1})D(\{G_k\}^n_{k=1})^{-1}\tilde{L}(\{G_k\}^n_{k=1}) = D^{-1}_{G^*_1} \cdots D^{-1}_{G^*_n}, \tag{4.39}$$

which follows easily by induction. ■

Theorem 4.32 The orthogonal polynomials satisfy the following Christoffel–Darboux-type formulas

$$(1-\bar{z}w)\sum_{k=0}^{n}\phi_k(z)^*\phi_k(w)=\hat{\phi}_{n+1}^{\mathrm{T}}(z)^*\hat{\phi}_{n+1}^{\mathrm{T}}(w)-\phi_{n+1}(z)^*\phi_{n+1}(w)$$

$$(1-z\bar{w})\sum_{k=0}^{n}\hat{\phi}_k(z)\hat{\phi}_k(w)^*=\phi_{n+1}^{\mathrm{T}}(z)\phi_{n+1}^{\mathrm{T}}(w)^*-\hat{\phi}_{n+1}(z)\hat{\phi}_{n+1}(w)^*.$$

Proof Using the recurrence relations in Theorem 4.31, we deduce

$$\hat{\phi}_{k+1}^{\mathrm{T}}(z)^*\hat{\phi}_{k+1}^{\mathrm{T}}(w)-\phi_{k+1}(z)^*\phi_{k+1}(w)=\hat{\phi}_k^{\mathrm{T}}(z)^*\hat{\phi}_k^{\mathrm{T}}(w)-\bar{z}w\phi_k(z)^*\phi_k(w)$$

and, adding up for $k=0,\cdots,n$, we obtain the first relation in the statement. In the same way we obtain the second one. ∎

Corollary 4.33 $\phi_n(z)$ and $\hat{\phi}_n(z)$ are invertible operators for $|z|\geq 1$.

Proof For the proof it is convenient to introduce the polynomials

$$\mathring{\Phi}_n(z)=D_{G_n}^{-1}\cdots D_{G_1}^{-1}\Phi_n(z) \tag{4.40}$$

$$\mathring{\hat{\Phi}}_n(z)=\hat{\Phi}_n(z)D_{G_1^*}^{-1}\cdots D_{G_n^*}^{-1}. \tag{4.41}$$

By Theorem 4.31 we find that these polynomials satisfy the recurrence formulas

$$D_{G_{n+1}}\mathring{\Phi}_{n+1}(z)=z\mathring{\Phi}_n(z)-G_{n+1}^*\mathring{\hat{\Phi}}_n^{\mathrm{T}}(z) \tag{4.42}$$

$$\mathring{\hat{\Phi}}_{n+1}(z)D_{G_{n+1}^*}=z\mathring{\hat{\Phi}}_n(z)-\mathring{\Phi}_n^{\mathrm{T}}(z)G_{n+1}^* \tag{4.43}$$

and by Theorem 4.32, they satisfy the Christoffel–Darboux type formulas:

$$(1-\bar{z}w)\sum_{k=0}^{n}\mathring{\Phi}_k(z)^*\mathring{\Phi}_k(w)=\mathring{\hat{\Phi}}_{n+1}^{\mathrm{T}}(z)^*\mathring{\hat{\Phi}}_{n+1}^{\mathrm{T}}(w)-\mathring{\Phi}_{n+1}(z)^*\mathring{\Phi}_{n+1}(w) \tag{4.44}$$

$$(1-z\bar{w})\sum_{k=0}^{n}\mathring{\hat{\Phi}}_k(z)\mathring{\hat{\Phi}}_k(w)^*=\mathring{\Phi}_{n+1}^{\mathrm{T}}(z)\mathring{\Phi}_{n+1}^{\mathrm{T}}(w)^*-\mathring{\hat{\Phi}}_{n+1}(z)\mathring{\hat{\Phi}}_{n+1}(w)^*. \tag{4.45}$$

Using (4.44) for $z=w$, we deduce that

$$\mathring{\Phi}_n^{*-1}(z)\mathring{\hat{\Phi}}_n^{\mathrm{T}}(z)\mathring{\hat{\Phi}}_n^{\mathrm{T}}(z)\mathring{\Phi}_n^{-1}(z)\leq I,\quad n\geq 0,|z|\geq 1. \tag{4.46}$$

Now, it is obvious that $\mathring{\Phi}_1(z)$ is invertible for $|z| \geq 1$ (we recall that we have assumed the hypothesis $\| G_n \| < 1, n \geq 1$). Suppose $\mathring{\Phi}_n(z)$ is invertible for $|z| \geq 1$. From (4.42),

$$\begin{aligned} D_{G_{n+1}} \mathring{\Phi}_{n+1}(z) &= z\mathring{\Phi}_n(z) - G_{n+1} \hat{\mathring{\Phi}}_n^{\mathrm{T}}(z) \\ &= (z - G_{n+1}^* \hat{\mathring{\Phi}}_n^{\mathrm{T}}(z) \mathring{\Phi}_n^{-1}(z)) \mathring{\Phi}_n(z). \end{aligned}$$

From (4.46) and the fact that $\| G_{n+1} \| < 1$ we conclude that $\mathring{\Phi}_{n+1}(z)$ is also invertible for $|z| \geq 1$. Having in mind the connection between $\mathring{\Phi}_n$, Φ_n and ϕ_n we get the required result. ∎

Corollary 4.34 $\phi_n^{\mathrm{T}}(z)$ and $\hat{\Phi}_n^{\mathrm{T}}(z)$ are invertible operators for $|z| \leq 1$.

We now introduce the polynomials of the second kind (Geronimus polynomials) by using Remark 4.11. Define

$$\begin{aligned} &P_s(e^{it}) = \frac{e^{it} + z}{e^{it} - z} \\ &\psi_0(z) = I \\ &\psi_n(z) = (P_z(\phi_n - \phi_n(z)), I)_{1\mu} \end{aligned} \tag{4.47}$$

and

$$\Psi_n(z) = (D_{G_1} \cdots D_{G_n}^2 \cdots D_{G_1})^{\frac{1}{2}} \psi_n(z); \tag{4.48}$$

then,

$$\begin{aligned} &\hat{\psi}_o(z) = I \\ &\hat{\psi}_n(z) = (P_z, \hat{\phi}_n - \hat{\phi}_n(z))_{r\mu} \end{aligned} \tag{4.49}$$

and

$$\hat{\Psi}_n(z) = \hat{\psi}_n(z)(D_{G_1^*} \cdots D_{G_n^*}^2 \cdots D_{G_1^*})^{\frac{1}{2}}. \tag{4.50}$$

Theorem 4.35 For $n \geq 1$,

$$\begin{aligned} \psi_n(z) &= \phi_n(z, -G_1, \cdots, -G_n) \\ \hat{\psi}_n(z) &= \hat{\phi}_n(z, -G_1, \cdots, -G_n). \end{aligned}$$

Proof From the definitions one deduces:

$$\Psi_{n+1}(z) - z\Psi_n(z) - D_{G_1} \cdots D_{G_n} G^*_{n+1} D^{-1}_{G^*_n} \cdots D^{-1}_{G^*_1} \hat{\Psi}^T_n(z)$$
$$= \frac{1}{2\pi} \int_0^{2\pi} \frac{e^{it}+z}{e^{it}-z} ((e^{it}-z)\Phi_n(e^{it}) - D_{G_1} \cdots D_{G_n} G^*_{n+1} D^{-1}_{G^*_1} \cdots D^{-1}_{G^*_1}$$
$$(e^{int} - z^n)\hat{\Phi}_n(e^{it}))^*) d\mu(t)$$

and taking into account the orthogonality relations $(\hat{\Phi}_n, \chi_k)_{r\mu} = 0,\ 0 \leq k < n$, it follows that the above integral is zero. The resulting recurrence formulas show exactly the required connections. ∎

Now, we define the polynomials

$$\mathcal{A}_n(z) = \frac{1}{2z}(\Psi^T_n(z) - \Phi^T_n(z)) \tag{4.51}$$

$$\mathcal{B}_n(z) = \frac{1}{2}(\Psi^T_n(z) + \Phi^T_n(z)) \tag{4.52}$$

$$\hat{\mathcal{A}}_n(z) = \frac{1}{2z}(\hat{\Psi}^T_n(z) - \hat{\Phi}^T_n(z)) \tag{4.53}$$

$$\hat{\mathcal{B}}_n(z) = \frac{1}{2}(\hat{\Psi}^T_n(z) + \hat{\Phi}^T_n(z)). \tag{4.54}$$

Moreover, consider F, the function in $\mathcal{S}(\mathcal{H})$ given by (4.15). We recall that its Schur parameters are exactly $\{G_n\}_{n\geq 1}$. Applying the Schur algorithm in Section 1.3, we also obtain the functions a_n, b_n, c_n, d_n by (1.39). The connection between these functions and the orthogonal polynomials (assuming of coures that they exist) is the following.

Theorem 4.36 For $n \geq 0$,

$$a_n(z) = \mathcal{A}_n(z)\mathcal{B}_n(z)^{-1} = \hat{\mathcal{B}}_n(z)^{-1}\hat{\mathcal{A}}_n(z)$$
$$b_n(z) = z^n \hat{\mathcal{B}}_n(z) D_{G^*_1} \cdots D_{G^*_n}$$
$$c_n(z) = D_{G_n} \cdots D_{G_1} \mathcal{B}_n(z)^{-1}$$
$$d_n(z) = -z D_{G_n} \cdots D_{G_1} \mathcal{B}_n(z)^{-1} \hat{\mathcal{A}}_n(z) D^{-1}_{G^*_1} \cdots D^{-1}_{G^*_n}$$

Proof By induction on n. ∎

We end this section with the following operatorial variant of Theorem 4.14 (1).

Theorem 4.37 $\{(\phi_n^{\mathrm{T}})^{-1}\}_{n\geq 0}$ converges uniformly on compact subsets of $\mathbf{D}$ in the strong operatorial topology to G_μ, the spectral factor of μ.

Proof We have, with the notation in Section 3.3, and Theorem 4.36,

$$\begin{aligned}\phi_n^{\mathrm{T}}(z)^{-1} = H_n' \Phi_n^{\mathrm{T}}(z)^{-1} &= H_n'(\mathcal{B}_n - z\mathcal{A}_n)^{-1} \\ &= H_n' D_{G_n}^{-1} \cdots D_{G_1}^{-1} c_n(z)(I - z a_n(z))^{-1} \\ &= \tilde{\mathcal{B}}_n' c_n(z)(I - z a_n(z))^{-1}\end{aligned}$$

and now the result follows from Theorem 1.12, Theorem 3.19 and Proposition 3.13. ∎

As a consequence of this result, we can obtain an operatorial version of Theorem 4.15. More precisely, we have the following formulas for computing the error prediction operator

$$\alpha = \inf\{\frac{1}{2\pi}\int_0^{2\pi} \mid p(e^{it}) \mid^2 d\mu(t) \mid p \in P \text{ and } p(0) = I\}$$

even when μ is a $\mathcal{L}(\mathcal{H})$–valued semispectral measure.

Corollary 4.38 The following formulas for computing hold true:

$$\alpha = s - \lim_{n\to\infty} D_{G_1^*} D_{G_2^*} \cdots D_{G_{n-1}^*} D_{G_n^*}^2 D_{G_{n-1}^*} \cdots D_{G_2^*} D_{G_1^*}$$

and

$$\alpha = G_\mu(0)^* G_\mu(0).$$

Notes

The results in Section 4.1 were developed in [83], but we included some more recent points in view connected with Levinson's algorithm and its version given in [17] – see Remark 4.3. For all this material see also [48]. Theorem 4.8 and Remark 4.9 are in [22] and [55] and Theorem 4.10 is a classical result in [3]. Szegö's theory in Section 4.2 is developed in [83] – see also [45] and [48].

Proposition 4.16 is a variant of a result in [8]. The Erdös–Turán condition appeared in [42] in connection with attempts to extend Szegö's theory to other classes of orthogonal polynomials. The two main results concerning these ideas are Theorem 4.22, which can be found in [72] and [73] and Theorem 4.23 proved in [65]. The proof of Theorem 4.22 presented here is taken from [64].

Szegö's limit theorems are given by Szegö in [80] and [81]. For an account involving the strong Szegö limit theorem with the solution of some problems concerning Ising models, see [82]. The version indicated in Theorem 4.29 is taken from [50]. The versions involving the choice parameters are indicated in [13] for the scalar case and in [22] for the matrix case. Other generalisations of the first Szegö limit theorem appear in [62]. There is a whole book, [15], devoted to these topics.

Matrix–valued orthogonal polynomials were systematically investigated quite recently – see, for instance, the paper [34]. The operatorial version presented in Section 4 was sketched in [27]. The first formula in Corollary 4.38 is an operatorial version (obtained in [23]) of a classical formula of Verblunsky (see [48]), while the second one is an operatorial version (obtained in [74] and [79]) of a classical formula of Szegö.

CHAPTER 5

A few 'electrical engineering' applications

First we have to explain the title. We intend to describe some other applications of the Schur algorithm to questions which are abstract formulations of some problems having their roots in electrical engineering, but it is not our purpose to describe them in detail. Thus, we present some elements of realisation theory (because this is a natural setting of Theorem 3.14 which we have already used, Darlington synthesis (because it is one more application of spectral factorisation) and stabilising control (because it may be a possible introduction for the Nehari problem).

5.1 Realisation theory

We return to the questions briefly discussed in Section 1.4. By a (time–invariant, with discrete time) linear system, we mean that there are given the equations:

$$\begin{cases} x_{n+1} = Ax_n + Bu_n \\ y_n \quad\ = Cx_n + Du_n \end{cases} \tag{5.1}$$

where $A \in \mathcal{L}(\mathcal{H})$, $B \in \mathcal{L}(\mathcal{U}, \mathcal{H})$ $C \in \mathcal{L}(\mathcal{H}, \mathcal{Y})$, $D \in \mathcal{L}(\mathcal{U}, \mathcal{Y})$, and the state space, $\mathcal{H}$, the input space, $\mathcal{U}$, and the output space, $\mathcal{Y}$, are separable Hilbert spaces. This description is an internal one of a one–port which evolves in time. Of course, this description is not realistic, because it is not always possible to know the content of a system, i.e. the space $\mathcal{H}$ and the operator A. Usually, we know only the answers to some questions posed to the system, i.e. an external description.

To be more precise, let us put the system to work; we get

$$y_n = CA^n x_o + CA^{n-1}Bu_o + CA^{n-2}Bu_1 + \cdots + CBu_{n-1} + Du_n$$

so that the function

$$F(z) = D + zC(I - zA)^{-1}B$$

is the transfer function of the given system.

Several formulations of the realisation problem can be stayed, and we consider the following one:

'Given $F \in \mathcal{S}(\mathcal{U}, \mathcal{Y})$, it is required to find a separable Hilbert space $\mathcal{H}$ and the operators $A \in \mathcal{L}(\mathcal{H})$, $B \in \mathcal{L}(\mathcal{U}, \mathcal{H})$, $C \in \mathcal{L}(\mathcal{H}, \mathcal{Y})$, $D \in \mathcal{L}(\mathcal{U}, \mathcal{Y})$ such that

$$F(z) = D + zC(I - zA)^{-1}B$$

and the node (or two–port)

$$S = \begin{bmatrix} A & B \\ C & D \end{bmatrix} : \begin{matrix} H \\ \oplus \\ \mathcal{U} \end{matrix} \rightarrow \begin{matrix} \mathcal{H} \\ \oplus \\ \mathcal{Y} \end{matrix}$$

is unitary.'

It is usual to say that F is also the transfer (or, even characteristic) function of the node S and that S is a realisation of F. The main result concerning this problem is the following.

Theorem 5.1 Let $F \in \mathcal{S}(\mathcal{U}, \mathcal{Y})$, $\{G_n\}_{n=1}^{\infty}$ be the Schur parameters of F and $W(\{G_n\}_{n=1}^{\infty})$ be the unitary operator associated by (2.35) to these parameters. Define, with the usual notation,

$$\mathcal{H} = \cdots \bigoplus \mathcal{D}_*(L) \bigoplus \mathcal{D}_*(L) \bigoplus \mathcal{D}_{G_1} \bigoplus \mathcal{D}_{G_1} \bigoplus \cdots$$

and

$$\begin{aligned} D &= G_1 \\ B &= (\cdots 0, D_{G_1}, 0, \cdots)^t \\ C &= (\cdots, H_\infty(L), D_{G_1^*} G_2 D_{G_1^*}, D_{G_2^*} G_3, \cdots) \\ A &= \text{What remains in the matrix of } W \text{ after deleting } D, B, C. \end{aligned}$$

Then,

$$S = \begin{bmatrix} A & B \\ C & D \end{bmatrix}$$

is a realisation of F. Moreover, if we define $\mathcal{C} = \bigvee_{n=0}^{\infty} A^n B \mathcal{U}$ and $\mathcal{O} = \bigvee_{n=0}^{\infty} A^{*n} C^* \mathcal{Y}$, then $\mathcal{C} \bigvee \mathcal{O} = \mathcal{H}$.

Proof Suppose first that $\mathcal{U} = \mathcal{Y}$. Further, since $\{G_n\}_{n=1}^{\infty}$ are the Schur parameters of F, let μ be the semispectral measure associated to these parameters by Theorem 2.14.

By Theorem 2.23, the operator $W = W(\{G_n\}_{n=1}^{\infty})$ associated to these parameters by (2.35) is exactly the Naimark dilation of μ. Let G be the function

$$G(z) = \frac{1}{2\pi} \int_0^{2\pi} \frac{e^{it} + z}{e^{it} - z} d\mu(t) \tag{5.2}$$

and then, by Proposition 2.18,

$$F(z) = (G(z) - I)(z(G(z) + I))^{-1}. \tag{5.3}$$

Moreover, from the properties of the Naimark dilation,

$$G(z) = P_{\mathcal{U}}^{\mathcal{K}}(I + zW)(I - zW)^{-1} \mid \mathcal{U} \tag{5.4}$$

($\mathcal{K}$ is the space where W is acting). By (5.3) and (5.4),

$$F(z) = (P_{\mathcal{U}}^{\mathcal{K}} W(I - zW)^{-1} \mid \mathcal{U})(P_{\mathcal{U}}^{\mathcal{K}}(1 - zW)^{-1} \mid \mathcal{U})^{-1}. \tag{5.5}$$

Using Proposition A.3, we can continue in (5.5) with

$$\begin{aligned} F(z) = (P_{\mathcal{U}}^{\mathcal{K}} \begin{bmatrix} A & B \\ C & D \end{bmatrix} \begin{bmatrix} * & z(1 - zA)^{-1}B(I - z(D + zC(1 - zA)^{-1}B))^{-1} \\ * & (I - z(D + zC(I - zA)^{-1}B)^{-1} \end{bmatrix} \mid \mathcal{U}) \\ \cdot (I - z(D + zC(I - zA)^{-1}B)) \\ = D + zC(I - zA)^{-1}B \end{aligned}$$

which is exactly the required formula.

For the general case of different $\mathcal{U}$ and $\mathcal{Y}$ we can proceed as in Remark 3.17. The property $\mathcal{C} \bigvee \mathcal{O} = \mathcal{H}$ follows easily from the structure of W. ■

Remark 5.2 Starting with the operator W given by (3.17) we derive as above the formula in Theorem 3.14.

The problem considered here is not the only one which can be formulated in this area. For instance, for engineering reasons, it can be asked whether, if the dimensions of $\mathcal{U}$ and $\mathcal{Y}$ are finite, the state space of the realisation will also be finite dimensional.

5.2 Darlington synthesis

The problem we are interested in is the following (with notation from Section 1.4).

'It is required to find the set of functions $F \in \mathcal{S}(\mathcal{U},\mathcal{Y})$ with the property that there exists a function $U \in \mathcal{S}(\mathcal{U} \bigoplus \mathcal{H}_1, \mathcal{Y} \bigoplus \mathcal{H}_2)$ inner and $*$–inner satisfying $F = C_U(0)$.'

In other words, it is required to find functions $B \in \mathcal{S}(\mathcal{H}_1,\mathcal{Y})$, $C \in \mathcal{S}(\mathcal{U},\mathcal{H}_2)$, $D \in \mathcal{S}(\mathcal{H}_1,\mathcal{H}_2)$ such that

$$U = \begin{bmatrix} F & B \\ C & D \end{bmatrix}$$

is inner from both sides. From the very beginning this problem involves the factorability of the functions $I - F^*(e^{it})F(e^{it})$ and $I - F(e^{it})F^*(e^{it})$ and not every function will have a Darlington synthesis. A general result which can be obtained in this direction is the following.

Theorem 5.3 If $F \in \mathcal{S}(\mathcal{U},\mathcal{Y})$, then for any compact K in $\mathbf{D}$ and $\epsilon > 0$, there exists $A \in \mathcal{S}(\mathcal{U},\mathcal{Y})$ such that

$$\sup_{z\in K} \| F(z) - A(z) \| < \epsilon$$

and A has a Darlington synthesis.

Proof This is a consequence of Theorem 1.12, namely that the functions a_n, $n \geq 0$, defined by (1.39) converge uniformly on compact subsets of $\mathbf{D}$ in the uniform operatorial topology to F and the fact that the functions defined by (1.37) are inner from both sides. ∎

Regarding the exact Darlington synthesis, we have the following result for the class $\mathcal{P}L^\infty(\mathcal{L}(\mathcal{U}))$ defined in Section 3.4.

Theorem 5.4 If $F \in \mathcal{P}L^\infty(\mathcal{L}(\mathcal{U})) \cap \mathcal{S}(\mathcal{U})$, then F has a Darlington synthesis.

Proof Consider the function

$$N(t) = I - F(e^{it})F^*(e^{it})$$

and let ϕ be the scalar inner function with the property that $\phi F^* \in H^\infty(\mathcal{L}(\mathcal{U}))$. Consequently, $\phi N = \phi I - (F)(\phi F^*) \in H^\infty(\mathcal{L}(\mathcal{U}))$, i.e. $N \in \mathcal{P}L^\infty(\mathcal{L}(\mathcal{U}))$ and, by Proposition 3.24 and Remark 3.25, N is factorable by the right spectral factor $R_F \in \mathcal{S}(\mathcal{H}_1,\mathcal{U})$. Consider now the function

$$F_1(z) : \mathcal{U} \bigoplus \mathcal{H}_1 \to \mathcal{U}$$
$$F_1(z) = (F(z), R_F(z))$$

which belongs to $\mathcal{S}(\mathcal{U} \bigoplus \mathcal{H}_1, \mathcal{U})$. By Remark 3.26,

$$\phi F_1^* = (\phi F^*, \phi R_F^*)^{\mathrm{t}} \in H^\infty(\mathcal{U}, \mathcal{U} \bigoplus \mathcal{H}_1)$$

so that $F_1 \in \mathcal{P}L^\infty(\mathcal{L}(\mathcal{U} \bigoplus \mathcal{H}_1, \mathcal{U}))$. (The definition of $\mathcal{P}L^\infty(\mathcal{U} \bigoplus \mathcal{H}_1, \mathcal{U})$ is the same as in the case $\mathcal{H}_1 = 0$, and all the properties in this case are inherited.) This means that $I - F_1^* F_1$ is also factorable so let $L_{F_1} \in \mathcal{S}(\mathcal{U} \bigoplus \mathcal{H}_1, \mathcal{H}_2)$ be its left spectral factor. Finally the function

$$U(z) = \mathcal{U} \bigoplus \mathcal{H}_1 \to \mathcal{U} \bigoplus \mathcal{H}_2$$
$$U(z) = (F_1(z), L_{F_1}(z))^{\mathrm{t}}.$$

has the required properties. It is obvious that $U^*U = I$ a.e. on $\mathbf{T}$ and let us show that $UU^* = I$ a.e. For this purpose, let us denote $L_{F_1} = (C, D)$. By direct computations,

$$UU^* = \begin{bmatrix} I & FC^* + R_F D^* \\ CF^* + DR_F^* & CC^* + DD^* \end{bmatrix} \quad \text{a.e. on } \mathbf{T}$$

and, as $UU^* \leq I$, it follows by Theorem A.7 that $FC^* + R_F D^* = 0$ a.e. on $\mathbf{T}$.

With these relations and with those resulting from the equality $U^*U = I$ a.e. on $\mathbf{T}$, we obtain

$$C^*(DD^* + CC^*) = C^*$$
$$D^*(DD^* + CC^*) = D^* \quad \text{a.e. on } \mathbf{T}$$

and in view of the outerness of L_{F_1}, $L_{F_1}^*(e^{it})$ is injective a.e. on $\mathbf{T}$, so that $DD^* + CC^* = I$ a.e. ∎

Let us return now to the class of functions considered in Remark 3.29. Denote by $P[\mathbf{C}]$ the set of polynomials with complex coefficients and by $R[\mathbf{C}]$ the set of rational functions with complex coefficients. Then, let $P^{n\times m}[\mathbf{C}]$ be the set of $n \times m$ matrices with elements in $P[\mathbf{C}]$ and $R^{n\times m}[\mathbf{C}]$ the set of $n \times m$ matrices with elements in $R[\mathbf{C}]$, where for simplicity we suppose that the poles of the functions in $R^{n\times m}[\mathbf{C}]$ are outside $\bar{\mathbf{D}}$. In this way, by Remark 3.29, $R^{n\times m}[\mathbf{C}] \subset \mathcal{P}L^\infty(\mathcal{L}(\mathbf{C}^m, \mathbf{C}^n))$.

Theorem 5.5 Let $Q \in R^{n\times m}[\mathbf{C}]$. Then $Q = A^{-1}B$, where $B \in P^{n\times m}[\mathbf{C}]$ and $A \in P^{n\times n}[\mathbf{C}]$ is non–singular in the sense that $\det A(z)$ is not identically zero. Moreover, there exist $\bar{X} \in P^{n\times n}[\mathbf{C}]$ and $\bar{Y} \in P^{m\times n}[\mathbf{C}]$ such that $A\bar{X} + B\bar{Y} = I$.

Proof As $Q \in \mathcal{P}L^\infty(\mathcal{L}(\mathbf{C}^m, \mathbf{C}^n))$, $Q = \frac{P}{d}$, with P a matrix and d a scalar polynomial with zeros outside $\bar{\mathbf{D}}$. By the theorem of invariant factors, there exist the matrix polynomials $U \in P^{(m+n)\times(m+n)}[\mathbf{C}]$ and $R \in P^{m\times m}[\mathbf{C}]$ such that

$$\begin{bmatrix} U_{11}(z) & U_{12}(z) \\ U_{21}(z) & U_{22}(z) \end{bmatrix} \begin{bmatrix} d(z)I_m \\ P(z) \end{bmatrix} = \begin{bmatrix} R(z) \\ 0 \end{bmatrix}$$

and $\det U(z)$ is a constant non–zero function. Define

$$V = U^{-1} = \begin{bmatrix} V_{11} & V_{12} \\ V_{21} & V_{22} \end{bmatrix} ;$$

then $d(z)I_m = V_{11}(z)R(z)$, so that V_{11} and R are non–singular polynomials, and by Proposition A.3 we conclude that U_{22} is also non–singular. Since

$$U_{21}(dI_m) + U_{22}P = 0$$

it follows that $Q = U_{22}^{-1}U_{21}$ and thus we take $A = U_{22}$ and $B = U_{21}$.

Finally, as $\det U$ is constant, V_{ij} are polynomials for $i, j = 1, 2$ and

$$U_{21}V_{12} + U_{22}V_{22} = I,$$

thus we take $\bar{X} = V_{22}$ and $\bar{Y} = V_{12}$. ■

Remark 5.6 In a similar way, we obtain, for $Q \in R^{n\times m}[\mathbf{C}]$, the representation $Q = CD^{-1}$ with certain polynomials C and D such that D is non–singular. There also exist the polynomials X and Y with

$$XD + YC = I.$$

We call a factorisation as in Theorem 5.8 a coprime left factorisation and the above one a coprime right factorisation.

These considerations provide a precise Darlington synthesis in the sense of the following result.

Theorem 5.7 For any coprime right factorisation CD^{-1} there exist polynomials, X, Y, X_*, Y_* and a coprime left factorisation $A^{-1}B$ such that

$$\begin{bmatrix} -X & Y \\ A & B \end{bmatrix} \begin{bmatrix} -C & X_* \\ D & Y_* \end{bmatrix} = I.$$

Proof By Remark 5.6, there exist polynomials X and Y such that

$$XC + YD = I.$$

By Theorem 5.5 for $Q = CD^{-1}$, there exists a coprime right factorisation $A^{-1}B$ of Q, so there exist polynomials $\bar{X}$, $\bar{Y}$ with

$$A\bar{X} + B\bar{Y} = I.$$

Consequently,

$$\begin{bmatrix} -X & Y \\ A & B \end{bmatrix} \begin{bmatrix} -C & \bar{X} \\ D & \bar{Y} \end{bmatrix} = \begin{bmatrix} I & T \\ 0 & I \end{bmatrix}$$

with $T = Y\bar{Y} - X\bar{X}$. But

$$\det \begin{bmatrix} I & T \\ 0 & I \end{bmatrix} = 1$$

and, with Proposition A.3, we get

$$\begin{bmatrix} -X & Y \\ A & B \end{bmatrix} \begin{bmatrix} -C & CT + \bar{X} \\ D & -DT + \bar{Y} \end{bmatrix} = \begin{bmatrix} I & 0 \\ 0 & I \end{bmatrix}.$$

■

5.3 Stabilising control

The functions in the class $R^{n \times m}[\mathbf{C}]$ are usually called stable rational functions and a system with a stable transfer function is also called stable. We consider the following problems.

' Given a linear system with transfer function F, find all the linear systems with transfer function G stabilising F, i.e. find all G such that the feedback system of Fig. 5.1 is stable.'

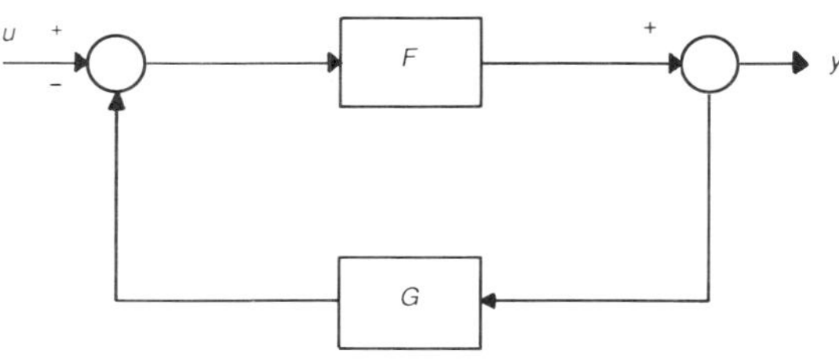

Fig. 5.1

The problem has the following solution.

Theorem 5.8 $G = (X_1 - KB_1)^{-1}(Y_1 + KA_1) = (Y + AK)(X - BK)^{-1}$, where K is the parameter belonging to the class of stable rational functions and the other elements are polynomials depending on F.

Proof First, the transfer function of the feedback system can be determined as follows:

$$y = FU - FGy$$
$$(I + FG)y = Fu$$
$$y = (I + FG)^{-1}Fu = F(I + GF)^{-1}u.$$

Then, use for F and G the coprime factorisations given by Theorem 5.8 and Remark 5.9, so

$$F = BA^{-1} \quad \text{and} \quad G = P^{-1}Q.$$

Consequently, the transfer function of the feedback system is

$$BA^{-1}(I + P^{-1}QBA^{-1})^{-1} = B(PA + QB)^{-1}P.$$

According to Theorem 5.7,

$$\begin{bmatrix} -Y & A \\ X & B \end{bmatrix} \begin{bmatrix} -B_1 & A_1 \\ X_1 & Y_1 \end{bmatrix} = \begin{bmatrix} I & 0 \\ 0 & I \end{bmatrix}$$

so that

$$(P,Q)\begin{bmatrix} -Y & A \\ X & B \end{bmatrix} = (P,Q)\begin{bmatrix} -B_1 & A_1 \\ X_1 & Y_1 \end{bmatrix}^{-1}$$

or

$$\left((P,Q)\begin{bmatrix} -Y \\ X \end{bmatrix}, (P,Q)\begin{bmatrix} A \\ B \end{bmatrix}\right) \begin{bmatrix} -B_1 & A_1 \\ X_1 & Y_1 \end{bmatrix} = (P,Q).$$

This can be rewritten as

$$(P,Q) = (N,D)\begin{bmatrix} -B_1 & A_1 \\ X_1 & Y_1 \end{bmatrix}$$

with a natural meaning for N and D. Finally

$$G = P^{-1}Q = (DX_1 - NB_1)^{-1}(DY_1 + NA_1) = (X_1 - D^{-1}NB_1)^{-1}(Y_1 + D^{-1}NA_1)$$

and taking $K = D^{-1}N$, we find that G is stable if and only if K is stable and the required formula holds. ∎

The final discussion offers a good opportunity for introducing the Nehari problem. Thus, to the feedback system considered above, add the noise u_2 and obtain the system illustrated in Fig. 5.2.

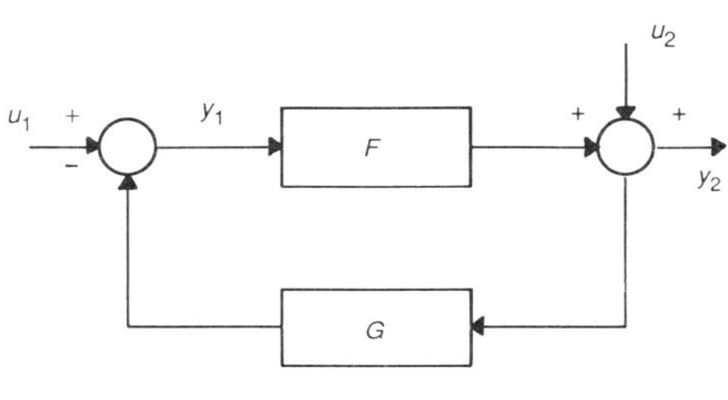

Fig. 5.2.

The transfer relations are:

$$\begin{bmatrix} y_1 \\ y_2 \end{bmatrix} = \begin{bmatrix} (I+GF)^{-1} & -G(I+FG)^{-1} \\ F(I+GF)^{-1} & (I+FG)^{-1} \end{bmatrix} \begin{bmatrix} u_1 \\ u_2 \end{bmatrix}.$$

We consider the direct relation between y_2 and u_2, $y_2 = (I+FG)^{-1}u_2$. $S = (I+FG)^{-1}$ is called the sensitivity of the feedback system because it expresses the dependence of the output on the noise. We consider also two filters to u_2 and y_2 and the weighted sensitivity $X(G) = W_1 S W_2$, where W_1, W_2 are non–singular, stable and with stable inverses.

The problem is the following.

'It is required to compute

$$\inf\{\| X(G) \|_{H^\infty} | \quad G \text{ stabilising feedback}\}.'$$

Using Theorem 5.8, we get

$$X(G) = W_1(I + A_1^{-1}B_1(Y+AK)(X-BK)^{-1})^{-1}W_2 = W_1(X-BK)A_1W_2$$

and we have to determine

$$\inf\{\| W_1(X - BK)A_1W_2 \|_{H^\infty} |\ K \text{ stable }\}.$$

Take the inner–outer factorisation of W_1B (according to Proposition 3.12), $W_1B = (W_1B)_i(W_1B)_o$ and the factorisation $A_1W_2 = (A_1W_2)_o(A_1W_2)_i$. Define

$$\begin{aligned}
\lambda &= \det(A_1W_2)_i \\
L &= W_1XA_1W_2\lambda(A_1W_2)_i^{-1} \\
U &= \lambda(W_1B)_i \\
H &= (W_1B)_oK(A_1W_2)_o
\end{aligned}$$

and obtain $X(G)\lambda(A_1W_2)_i^{-1} = L - UH$. Since $\lambda(A_1W_2)_i^{-1}$ is inner, the problem is reduced to the computation of

$$\inf\{\| L - UH \|_{H^\infty} |\ H \text{ stable }\}, U \text{ inner}$$

which will be solved in the next chapter.

Notes

From a very large literature we can indicate only a few references. Thus, Theorem 5.1 was proved in [54] and detailed treatment of the realisation theory can be found in [56] and [52]. A recent paper where the continued fraction decomposition of linear systems was considered is [5]. Concerning partial realisation we indicate [47].

For the Darlington synthesis we have used references [33]. [7] and [38] and for matrix polynomials and rational fractions we used [52]. Theorem 5.11 is a result in [91] and the approach to control problems is explained in Section 5.3 appears in [89]; here we used [44].

CHAPTER 6

Other completion problems

Another method for solving completion problems is based on the extension of partial isometries to unitary operators. Our purpose in this chapter is to connect this method with the elements involved by the Schur algorithm. Some other problems will be taken into account, including Nevanlinna–Pick's. In addition, the elements of lifting of commutants will be treated in this context.

6.1 The main problem

The following problem constitutes our main concern.

'Given a partial isometry $w \in \mathcal{L}(\mathcal{H})$, acting on a given Hilbert space $\mathcal{H}$ it is required to find all minimal unitary extensions of w.'

By a minimal unitary extension of w we understand here a unitary operator W acting on a Hilbert space $\mathcal{K}$ containing $\mathcal{H}$, such that $Wh = wh$ for $h \in \mathcal{H}$ and $\bigvee_{n\in\mathbf{Z}} W^n\mathcal{H} = \mathcal{K}$. Of course, a natural equivalence is imposed on the set of minimal unitary extensions of w. Two such unitary operators $W \in \mathcal{L}(\mathcal{K})$ and $W' \in \mathcal{L}(\mathcal{K}')$ are equivalent if there exists a unitary operator $\phi \in \mathcal{L}(\mathcal{K},\mathcal{K}')$ with $\phi \mid \mathcal{H} = I$ and $W'\phi = \phi W$. We denote by $E(w)$ the classes of equivalent minimal unitary extensions of w, and this is the object we are interested in. Denote by $\mathcal{E}$ the initial space of w, $\mathcal{E} = \mathcal{H} \ominus \ker w$ and by $\mathcal{F}$ the final space of w, $\mathcal{F} = \mathcal{R}(w)$. Moreover, take $\mathcal{M} = \mathcal{H} \ominus \mathcal{E}$ and $\mathcal{N} = \mathcal{H} \ominus \mathcal{F}$. We can prove the first result.

Theorem 6.1 There exists a one–to–one correspondence between $E(w)$ and the set π of families of contractions $\{G_n\}_{n=1}^{\infty}$, $G_1 \in \mathcal{L}(\mathcal{M},\mathcal{N})$, $G_n \in \mathcal{L}(\mathcal{D}_{G_{n-1}},\mathcal{D}_{G^*_{n-1}})$, $n \geq 2$.

Proof Consider $\mathcal{G} = \{G_n\}_{n=1}^{\infty} \in \pi$ and define the row–contraction $L = L(\{G_n\}_{n=1}^{\infty}) : \mathcal{M} \bigoplus \bigoplus_{n=1}^{\infty} \mathcal{D}_{G_n} \to \mathcal{N}$. Consider the spaces:

$$\mathcal{K}_1 = \mathcal{K}_1(\mathcal{G}) = \cdots \bigoplus \mathcal{D}_*(L) \bigoplus \cdots \bigoplus \mathcal{D}_*(L) \bigoplus \mathcal{M} \bigoplus \bigoplus_{n=1}^{\infty} \mathcal{D}_{G_n} \tag{6.1}$$

$$\mathcal{K}_2 = \mathcal{K}_2(\mathcal{G}) = \cdots \bigoplus \mathcal{D}_*(L) \bigoplus \cdots \bigoplus \mathcal{D}_*(L) \bigoplus \mathcal{N} \bigoplus \bigoplus_{n=1}^{\infty} \mathcal{D}_{G_n} \tag{6.2}$$

and the unitary operator

$$\begin{aligned} &W_o(\mathcal{G}) : \mathcal{K}_1 \to \mathcal{K}_2 \\ &W_o(\mathcal{G}) = I \bigoplus \begin{bmatrix} I & 0 \\ 0 & \alpha(L) \end{bmatrix} J(L) \begin{bmatrix} 0 & I \\ B^*(L) & 0 \end{bmatrix} \end{aligned} \tag{6.3}$$

which is written with respect to the decompositions

$$\mathcal{K}_1 = (\cdots \bigoplus \mathcal{D}_*(L)) \bigoplus (\mathcal{D}_*(L) \bigoplus \mathcal{M} \bigoplus \bigoplus_{n=1}^{\infty} \mathcal{D}_{G_n})$$

and

$$\mathcal{K}_2 = (\cdots \bigoplus \mathcal{D}_*(L)) \bigoplus (\mathcal{N} \bigoplus \bigoplus_{n=1}^{\infty} \mathcal{D}_{G_n})$$

respectively. Finally, define

$$\mathcal{K}(\mathcal{G}) = \mathcal{K}_1(\mathcal{G}) \bigoplus \mathcal{E} = \mathcal{K}_2(\mathcal{G}) \bigoplus \mathcal{F} \tag{6.4}$$

and the unitary operator

$$\begin{aligned} &W(\mathcal{G}) : \mathcal{K} \to \mathcal{K} \\ &W(\mathcal{G}) = W_o(\mathcal{G}) \bigoplus w. \end{aligned} \tag{6.5}$$

It is quite obvious now that $W(\mathcal{G})$ is a minimal unitary extension of w and we can define the map:

$$\begin{aligned} &\Phi : \pi \to E(w) \\ &\Phi(\mathcal{G}) = \text{ class } (W(\mathcal{G})). \end{aligned} \tag{6.16}$$

Now we have to show that Φ is a bijective map. Supposing that $\Phi(\mathcal{G}) = \Phi(\mathcal{G}')$, there exists a unitary operator $\phi \in \mathcal{L}(\mathcal{K}(\mathcal{G}), \mathcal{K}(\mathcal{G}'))$ with $\phi \mid \mathcal{H} = I$ and $W(\mathcal{G}')\phi = \phi W(\mathcal{G})$. For $n \geq 0$, $W^n(\mathcal{G}')\phi = \phi W^n(\mathcal{G})$ and then

$$\begin{aligned} P_{\mathcal{H}}^{\mathcal{K}(\mathcal{G})} W^n(\mathcal{G}) \mid \mathcal{H} &= P_{\mathcal{H}}^{\mathcal{K}(\mathcal{G})} W^n(\mathcal{G})\phi^* \mid \mathcal{H} \\ &= P_{\mathcal{H}}^{\mathcal{K}(\mathcal{G})} \phi^* W^n(\mathcal{G}') \mid \mathcal{H} = P_{\mathcal{H}}^{\mathcal{K}(\mathcal{G}')} W(\mathcal{G}') \mid \mathcal{H}. \end{aligned}$$

Taking into account Theorem 2.33 and Theorem 2.14, it follows that $G_n = G'_n$ for $n \geq 1$, i.e. $\mathcal{G} = \mathcal{G}'$.

Now, let an element in $E(w)$ be represented by a unitary extension $W \in \mathcal{L}(\mathcal{K})$. Defining $S_1 = P^{\mathcal{K}}_{\mathcal{H}} W \mid \mathcal{H}$, S_1 does not depend on the chosen representant and $S_1 = w \bigoplus G_1$ where $G_1 \in \mathcal{L}(\mathcal{M}, \mathcal{N})$ is a contraction. Further, define $S_n = P^{\mathcal{K}}_{\mathcal{H}} W^n \mid \mathcal{H}$, $n > 0$, $S_0 = I_{\mathcal{H}}$ and $S_{-n} = S_n^*$, $n > 0$. Then $\{S_n\}_{n \in \mathbf{Z}}$ are the coefficients of a positive definite Toeplitz kernel on $\mathbf{Z}$ and again, S_n do not depend on the chosen representant. Let $\{G'_n\}_{n=1}^{\infty}$ be the parameters associated by Theorem 2.14 to this Toeplitz kernel. Then,

$$G'_1 = w \bigoplus G_1$$

and

$$\mathcal{D}_{G'_1} = 0 \bigoplus \mathcal{D}_{G_1}, \quad \mathcal{D}_{G'^*_1} = 0 \bigoplus \mathcal{D}_{G_1^*}.$$

Consequently, $\mathcal{G} = \{G_1\} \cup \{G'_n\}_{n=2}^{\infty}$ belongs to π and it is clear that $\Phi(\mathcal{G}) =$ class W. ∎

Corollary 6.2 $E(w)$ contains a unique element if and only if $\mathcal{M} = \{0\}$ or $\mathcal{N} = \{0\}$.

Another way of parametrising the set $E(W)$ is based on the notion of the generalised resolvent. Take $W \in \mathcal{L}(\mathcal{K})$ and $W' \in \mathcal{L}(\mathcal{K}')$ to be two extensions of w belonging to the same class (i.e. determining the same element in $E(w)$). Then there exists a unitary operator $\phi \in \mathcal{L}(\mathcal{K}, \mathcal{K}')$ such that $\phi \mid \mathcal{H} = I$ and $W'\phi = \phi W$. Consequently,

$$\begin{aligned} P^{\mathcal{K}}_{\mathcal{H}}(I - zW)^{-1} \mid \mathcal{H} &= P^{\mathcal{K}}_{\mathcal{H}})(I - zW)^{-1}\phi^* \mid \mathcal{H} \\ &= P^{\mathcal{K}}_{\mathcal{H}}\phi^*(I - zW')^{-1} \mid \mathcal{H} = P^{\mathcal{K}'}_{\mathcal{H}}(I - zW')^{-1} \mid \mathcal{H}. \end{aligned}$$

This shows that we can define the generalised resolvent of an element in $E(w)$ by taking $R(z) = P^{\mathcal{K}}_{\mathcal{H}}(I - zW)^{-1} \mid \mathcal{H}$ for a certain representant of it. Moreover, as in the proof of Theorem 6.1 we exhibited the structure of a certain representant of each element in $E(w)$, i.e. $W(\mathcal{G})$ given by (6.5). We choose to work with this special representant and denote by $R(\mathcal{G}; z)$ the generalised resolvent of the element in $E(w)$ associated by Theorem 6.1 to the parameter $\mathcal{G} \in \pi$. One more notation, W^o, distinguishes the unitary extension associated to the parameter $\mathcal{G}^o = \{0, 0, \cdots\}$, i.e.$W^o = W(\mathcal{G}^o)$. We set $R^o(z) = R(\mathcal{G}^o; z)$.

Now, take $W = W(\mathcal{G})$ and decompose it with respect to the sum $\mathcal{K}(\mathcal{G}) = \mathcal{H} \oplus (\mathcal{K}(\mathcal{G}) \ominus \mathcal{H})$,

$$W = \begin{bmatrix} W_1 & \hat{C} \\ \hat{B} & \hat{A} \end{bmatrix}$$

By Proposition A.3

$$\begin{aligned}
R(\mathcal{G};z) &= P_{\mathcal{H}}^{\mathcal{K}} \begin{bmatrix} I - zW_1 & -z\hat{C} \\ -z\hat{B} & I - z\hat{A} \end{bmatrix}^{-1} | \mathcal{H} \\
&= (I - zW_1)^{-1} + z^2(I - zW_1)^{-1}\hat{C}(I - z\hat{A} - z^2\hat{B}(I - zW_1)^{-1}\hat{C})^{-1}\hat{B}(I - zW_1)^{-1} \\
&= (I - zW_1)^{-1} \\
&+ z^2(I - zW_1)^{-1}\hat{C}(I - z^2(I - z\hat{A})^{-1}\hat{B}(I - zW_1)^{-1}\hat{C})^{-1}(I - z\hat{A})^{-1}\hat{B}(I - zW_1)^{-1} \\
&= (I - zW_1)^{-1} \\
&+ z^2(I - zW_1)^{-1}(I - z^2\hat{C}(I - z\hat{A})^{-1}\hat{B}(I - zW_1)^{-1})^{-1}\hat{C}(I - z\hat{A})^{-1}\hat{B}(I - zW_1)^{-1} \\
&= (I - zW_1)^{-1} \\
&+ z^2(I - zW_1)^{-1}\hat{C}(I - z\hat{A})^{-1}\hat{B}(I - z^2(I - zW_1)^{-1}\hat{C}(I - z\hat{A})^{-1}\hat{B})^{-1}(I - zW_1)^{-1}.
\end{aligned}$$

Now, we define

$$\hat{D} = W_1 - W_1^o$$

and then

$$(I - zW_1)^{-1} - (I - zW_1^o)^{-1} = (I - zW_1^o)^{-1}\hat{D}(I - zW_1)^{-1}.$$

Using this relation, we get

$$\begin{aligned}
(I - zW_1)^{-1} &= (I - z(I - zW_1^o)^{-1}\hat{D})^{-1}(I - zW_1^o)^{-1} \\
&= (I - zW_1^o)^{-1}(I - z\hat{D}(I - zW_1^o)^{-1})^{-1}.
\end{aligned}$$

Consequently,

$$\begin{aligned}
(I - zW_1)^{-1} - (I - zW_1^o)^{-1} &= z(I - zW_1^0)^{-1}\hat{D}(I - z(I - zW_1^o)^{-1}\hat{D})^{-1}(I - zW_1^o)^{-1} \\
&= z(I - zW_1)^{-1}(I - z\hat{D}(I - zW_1)^{-1})^{-1}\hat{D}(I - zW_1)^{-1}.
\end{aligned}$$

Defining $\hat{F}(z) = \hat{D} + z\hat{C}(I - z\hat{A})^{-1}\hat{B}$ and observing that $R^o(z) = (I - zW_1^o)^{-1}$, we finally get

$$\begin{aligned}
R(\mathcal{G};z) - R^o(z) &= zR^o(z)\hat{D}(I - zR^o(z)\hat{D})^{-1}R^o(z) \\
&\quad + z^2R^o(z)(I - z\hat{D}R^o(z))^{-1}\hat{C}(I - z\hat{A})^{-1}\hat{B}(I - z^2(I - zR^o(z)\hat{D})^{-1}R^o(z) \\
&\quad \hat{C}(I - z\hat{A})^{-1}\hat{B}) - 1(I - zR^o(z)\hat{D})^{-1}R^o(z) \\
&= zR^o(z)(I - z\hat{D}R^o(z))^{-1}(\hat{D} + z\hat{C}(I - z\hat{A})^{-1}\hat{B}(I - zR^o(z)\hat{F}(z))^{-1}R^o(z) \\
&= zR^o(z)(I - z\hat{D}R^o(z))^{-1}(-z\hat{D}R^o(z)\hat{F}(z) + \hat{F}(z))(I - zR^o(z)\hat{F}(z))^{-1}R^o(z) \\
&= zR^o(z)\hat{F}(z)(I - zR^o(z)\hat{F}(z))^{-1}R^o(z).
\end{aligned}$$

Now, we have to remind ourselves of some of the elements involved in the realisation theory presented in Chapter 5. We define

$$\begin{aligned}
D &= G_1 \\
B &= (\cdots, 0, 0, D_{G_1}, 0, \cdots)^{\mathrm{t}} \\
C &= (\cdots, 0, H_\infty(L), D_{G_1^*} G_2, D_{G_1^*} D_{G_2^*} G_3, \cdots) \\
A &= \text{what remains in } W(\mathcal{G}) \text{ after deleting } D, B \text{ and } C
\end{aligned}$$

(here of course $H_\infty(L) = s - \lim_{n\to\infty} D_{G_1^*} \cdots D^2_{G_n^*} \cdots D_{G_1^*})^{\frac{1}{2}}$). Then, we simply observe that $\hat{A}, \hat{B}, \hat{C}, \hat{D}$ are extensions of A, B, C, D with 0 on corresponding spaces: consequently, the above derived formula for $R(\mathcal{G}; z)$ becomes the following generalised resolvent formula:

$$R(\mathcal{G}; z) = R^o(z) + zR^o(z)P^{\mathcal{K}}_{\mathcal{N}} F(z)(I - X(z)F(z))^{-1} P^{\mathcal{H}}_{\mathcal{M}} R^o(z) \tag{6.7}$$

where

$$F(z) = D + zC(I - zA)^{-1}B \tag{6.8}$$

and

$$X(z) = zP^{\mathcal{H}}_{\mathcal{M}}(I - zW_1^o)^{-1} \mid \mathcal{N}. \tag{6.9}$$

In other words, F is the characteristic function of the unitary node

$$\begin{bmatrix} A & B \\ C & D \end{bmatrix}$$

and X is the characteristic function of the unitary node

$$\begin{bmatrix} W_1^o & P^{\mathcal{H}}_{\mathcal{N}} \mid \mathcal{N} \\ P^{\mathcal{H}}_{\mathcal{M}} & 0 \end{bmatrix} : \begin{matrix} \mathcal{H} \\ \oplus \\ \mathcal{N} \end{matrix} \rightarrow \begin{matrix} \mathcal{H} \\ \oplus \\ \mathcal{M}. \end{matrix}$$

As a consequence $F \in \mathcal{S}(\mathcal{M}, \mathcal{N})$ and $X \in \mathcal{S}(\mathcal{N}, \mathcal{M})$. Now we can summarise the connections between the Schur algorithm, generalised resolvents, realisation theory and Naimark dilation in the following result.

Theorem 6.3 There exists a one–to–one correspondence between $E(w)$ and $\mathcal{S}(\mathcal{M}, \mathcal{N})$ realised by the generalised resolvent formula (6.7). Moreover, we have the following commutative diagram:

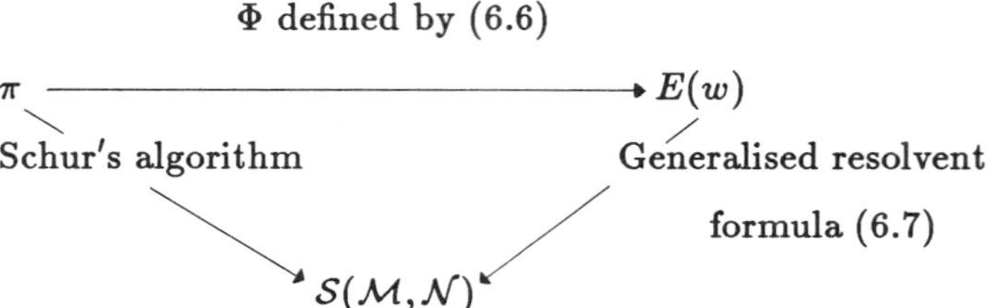

Proof The proof takes only a few lines. Define

$$\Phi_1 : \pi \to \mathcal{S}(\mathcal{M},\mathcal{N})$$
$$\Phi_1(\mathcal{G}) = \text{The function produced by Schur's algorithm} \tag{6.10}$$

and

$$\Phi_1 : E(w) \to \mathcal{S}(\mathcal{M},\mathcal{N})$$
$$\Phi_2(\text{class } W(\mathcal{G})) = F \text{ produced by (6.8) in the generalised resolvent formula} \tag{6.11}$$

and note that, by Theorem 5.1,

$$\Phi_1(\Phi(\mathcal{G})) = \Phi_1(\mathcal{G}), \quad \mathcal{G} \in \pi.$$ ■

It turns out that it is quite useful to consider, instead of the generalised resolvent, the following function:

$$S(\mathcal{G}; z) = P_{\mathcal{H}}^{\mathcal{K}(\mathcal{G})}(I + zW(\mathcal{G}))(I - zW(\mathcal{G}))^{-1} \mid \mathcal{H}.$$

Again, this function depends only on the class of $W(\mathcal{G})$ and it has the (obvious) distinguishing property that $S(\mathcal{G}; z) \in \mathcal{C}(\mathcal{H})$. Sometimes, $S(\mathcal{G}; z)$ is called the generalised coresolvent of the class of $W(\mathcal{G})$ and the appropriate generalised coresolvent formula can be deduced:

$$S(\mathcal{G}; z) = S(\mathcal{G}^o; z) + 2zR^o(z)P_{\mathcal{N}}^{\mathcal{H}}F(z)(I - X(z)F(z))^{-1}P_{\mathcal{M}}^{\mathcal{H}}R^o(z) \tag{6.12}$$

with the same definitions of the elements involved as in the case of the generalised resolvent formula (6.7). Indeed,

$$S(\mathcal{G};z) - S(\mathcal{G}^o;z) = 2R(\mathcal{G};z) - 2R(\mathcal{G}^o;z)$$

and (6.12) follows from (6.7). Using the cascade transformation defined Section 1.4, we can rewrite (6.12) in the form

$$S(\mathcal{G};\cdot) = C_{R_o}(F) \tag{6.13}$$

where

$$R_o(z) = \begin{bmatrix} S(\mathcal{G}^0;z) & \sqrt{2}zR^o(z)P^{\mathcal{H}}_{\mathcal{N}} \\ \sqrt{2}P^{\mathcal{H}}_{\mathcal{M}}R^o(z) & X(z) \end{bmatrix} : \begin{matrix} \mathcal{H} \\ \bigoplus \\ \mathcal{N} \end{matrix} \rightarrow \begin{matrix} \mathcal{H} \\ \bigoplus \\ \mathcal{M} \end{matrix} \tag{6.14}$$

is the so–called resolvent matrix of w.

We end this section by giving an example of the method for the case of the trigonometric moment problem as stated in its operatorial version in Section 2.4. Given the operators $S_k \in \mathcal{L}(\mathcal{H}_o)$, $k = 0,\cdots,n$ $(S_o = I)$, suppose that

$$M_n = \begin{bmatrix} I & S_1 & \cdots & S_n \\ S_1^* & I & \cdots & S_{n-1} \\ \cdot & & & \cdot \\ \cdot & & & \cdot \\ \cdot & & & \cdot \\ S_n^* & & \cdots & I \end{bmatrix}$$

is a positive block matrix.

Let us define the space $\mathcal{H}$ by renorming $\mathcal{H}_o \bigoplus \underset{n+1}{\cdots} \bigoplus \mathcal{H}_o$ with M_n, i.e. considering on $\mathcal{H}_o \bigoplus \underset{n+1}{\cdots} \bigoplus \mathcal{H}_o$ the inner product

$$(h,g)_{M_n} = (M_n h, g) \quad h,g \in \mathcal{H}_o \bigoplus \underset{n+1}{\cdots} \bigoplus \mathcal{H}_o,$$

factoring out its kernel if necessary. Furthermore, we consider the following subspaces of $\mathcal{H}$,

$$\mathcal{F} = \text{The subspace generated by } \{(0,h_1,h_2,\cdots,h_n) \mid h_k \in \mathcal{H}_o, k = 1,\cdots,n\}$$

and

$$\mathcal{E} = \text{The subspace generated by } \{(h_o,h_1,\cdots,h_{n-1},0) \mid h_k \in \mathcal{H}_o, k = 0,\cdots,n-1\}.$$

The operator $w : \mathcal{E} \to \mathcal{F}$, first defined by

$$w(h_o, h_1, \cdots, h_{n-1}, 0) = (0, h_o, \cdots, h_{n-1}),$$

extends then to a unitary operator from $\mathcal{E}$ onto $\mathcal{F}$, in view of the fact that M_n is a Toeplitz matrix. This extension is also denoted by w and, of course, w can be viewed as a partial isometry on $\mathcal{H}$ with initial space $\mathcal{E}$ and final space $\mathcal{F}$. A final step of the construction is to embed $\mathcal{H}_o$ into $\mathcal{H}$ by identifying h_o in $\mathcal{H}_o$ with the class of $(h_o, 0, 0, \cdots, 0)$ in $\mathcal{H}$.

Now in view of Theorem 6.1, $E(w) \neq \emptyset$ and take $W \in \mathcal{L}(\mathcal{K})$ to be an extension of w. Then the semispectral measure

$$\mu(\chi_p) = P^{\mathcal{K}}_{\mathcal{H}_o} W^{*p} \mid \mathcal{H}_o, \quad p \in \mathbf{Z},$$

is a solution of the trigonometric moment problem being considered. Indeed, for $0 \leq k \leq n$ (and with an obvious abuse of notation regarding the position of elements of $\mathcal{H}_o$ in $\mathcal{H}$),

$$\begin{aligned}
(P^{\mathcal{K}}_{\mathcal{H}_o} W^k h_o, g_o)_{M_n} &= (W^k h_o, g_o)_{M_n} = (W h_o, W^{*k-1} g_o)_{M_n} \\
&= (w h_o, W^{*k-1} g_o)_{M_n} = ((0, h_o, 0, \cdots, 0)^t, W^{*k-1} g_o)_{M_n} \\
&= (w(0, \cdots, 0, h_o, 0 \cdots, 0)^t, g_o)_{M_n} \\
&= ((0, \cdots, h_o, 0, \cdots, 0)^t, g_o)_{M_n} \\
&= (M_n(0, \cdots, h_n, \cdots, 0)^t, (g_o, 0, \cdots, 0)^t) = (S_k h_o, g_o) \\
&= (S_k h_o, g_o)_{M_n} \quad \text{for any } h_o, g_o \in \mathcal{H}_o.
\end{aligned}$$

Consequently

$$S_k = \mu(\chi_{-k}) \quad \text{for } 0 \leq k \leq n.$$

6.2 The Nehari problem

In this section we explore another classical interpolation problem. The solution is provided via the Hankel operators by the method developed in the previous section.

We consider the following formulation of the Nehari problem.

'For given operators $\{C_n\}_{n=-\infty}^{-1}$, $C_n \in \mathcal{L}(\mathcal{U})$, $\mathcal{U}$ being a separable Hilbert space, it is required to find conditions for the existence of a measurable function f on $\mathbf{T}$ with values that are contractions in $\mathcal{L}(\mathcal{U})$ such that

$$C_n = \frac{1}{2\pi}\int_0^{2\pi} e^{-int} f(e^{it})dt, \quad n = -1, -2, \cdots'$$

A solution of this problem is stated in terms of the Hankel operator associated to the sequence $\{C_n\}_{n=-\infty}^{-1}$. For this purpose, take $f \in L^\infty(\mathcal{U})$ and let

$$f_n = \frac{1}{2\pi}\int_0^{2\pi} e^{-int} f(e^{it})dt, \quad n \in \mathbf{Z},$$

be its Fourier coefficients. Denote by $\ell^2_+(\mathcal{U})$ the space of square summable $\mathcal{U}$–valued sequences indexed on $\{0, 1, 2, \cdots\}$ and by $\ell^2_-(\mathcal{U})$ the space of square summable $\mathcal{U}$–valued sequences indexed on $\{\cdots, -2, -1\}$. In this way, $\ell^2(\mathcal{U}) = \ell^2_-(\mathcal{U}) \bigoplus \ell^2_+(\mathcal{U})$, and the Fourier transform establishes an isomorphism between $\ell^2(\mathcal{U})$ and $L^2(\mathcal{U})$.

The multiplication operator is defined by the matrix:

$$L_f : \ell^2(\mathcal{U}) \to \ell^2(\mathcal{U})$$

$$L_f = \begin{bmatrix} & & & \cdot & & & \\ & & & \cdot & & & \\ & & & \cdot & & & \\ & \cdots & & f_{-2} & \cdots & & \\ \cdots & f_1 & f_o & f_{-1} & f_{-2} & \cdots & \\ & \cdots & & f_o & f_{-1} & f_{-2} & \cdots \\ & & & \cdot & & & \\ & & & \cdot & & & \\ & & & \cdot & & & \end{bmatrix} \tag{6.15}$$

We have that $\| L_f \| = \| f \|_\infty = \operatorname{ess\,sup}_{t\in[0,2\pi]} \| f(e^{it}) \|$ and if we write this matrix with respect to the decomposition $\ell^2(\mathcal{U}) = \ell^2_-(\mathcal{U}) \bigoplus \ell^2_+(\mathcal{U})$, then its corner acting between $\ell^2_+(\mathcal{U})$ and $\ell^2_-(\mathcal{U})$ is the Hankel operator associated to f and it is denoted by H_f. Returning to our problem, we also denote by $H(\{C_n\}_{n=-\infty}^{-1})$ the Hankel matrix based on $\{C_n\}_{n=-\infty}^{-1}$, i.e.

$$H(\{C_n\}_{n=-\infty}^{-1}) = \begin{bmatrix} \cdot & & & \\ \cdot & & & \\ \cdot & & & \\ C_{-3} & \cdots & & \\ C_{-1} & C_{-3} & \cdots & \\ C_{-1} & C_{-2} & C_{-3} & \cdots \end{bmatrix} : \ell^2_+(\mathcal{U}) \to \ell^2_-(\mathcal{U}). \tag{6.16}$$

Now we can obtain a solution of the Nehari problem.

Theorem 6.4 The Nehari problem is solvable if and only if $H(\{C_n\}_{n=-\infty}^{-1})$ is a contraction.

Proof The necessity is obvious, because if f is a solution of the problem, then $H(\{C_n\}_{n=-\infty}^{-1}) = H_f$, and

$$\| H_f \| = \| P_{\ell^2_-(\mathcal{U})}^{\ell^2(\mathcal{U})} L_f \mid \ell^2_+(\mathcal{U}) \| \leq \| L_f \| = \| f \|_\infty \leq 1.$$

For the converse, we consider $H = H(\{C_n\}_{n=-\infty}^{-1})$ to be a contraction, and then the matrix,

$$M = \begin{bmatrix} I & H \\ H^* & I \end{bmatrix} : \begin{matrix} \ell^2_-(\mathcal{U}) \\ \oplus \\ \ell^2_+(\mathcal{U}) \end{matrix} \to \begin{matrix} \ell^2_-(\mathcal{U}) \\ \oplus \\ \ell^2_+(\mathcal{U}) \end{matrix} \tag{6.17}$$

is positive by Theorem A.1. Renorming $\ell^2(\mathcal{U}) = \ell^2_-(\mathcal{U}) \bigoplus \ell^2_+(\mathcal{U})$ with M, we get a Hilbert space $\mathcal{H}$ and the following two subspects are also considered:

$$\begin{aligned} \mathcal{E} &= \text{The subspace of } \mathcal{H} \text{ generated by the set} \\ &\quad \{h = (h_n)_{n=-\infty}^{\infty} \mid h \text{ has finite support and } h_{-1} = 0\} \\ \mathcal{F} &= \text{The subspace of } \mathcal{H} \text{ generated by the set} \\ &\quad \{h = (h_n)_{n=-\infty}^{\infty} \mid h \text{ has finite support and } h_o = 0\}. \end{aligned}$$

Moreover $\mathcal{U}$ is regarded as sitting isometrically inside $\mathcal{H}$, by simply considering $\mathcal{U} \ni h \to (\cdots 0, h, 0, \cdots)$, with h on the position zero. The operator $w : \mathcal{E} \to \mathcal{F}$, defined by

$$w(\cdots, h_{-2}, 0, \boxed{h_o}, h_1, \cdots) = (\cdots, h_{-3,}, h_{-2}, \boxed{0}, h_o, h_1, \cdots)$$

for sequences with finite support, extends to a unitary operator from $\mathcal{E}$ onto $\mathcal{F}$, in view of the fact that H is a Hankel matrix. Indeed, considering $h^{(j)} = (\cdots 0, h, 0, \cdots)$ with the element on $h \in \mathcal{U}$ on the position j, we have

$$(h^{(j)}, g^{(k)})_{\mathcal{H}} = \begin{cases} (C_{k-j}h, g) & \text{if } j \geq 0 > k \\ \delta_{jk} & \text{elsewhere} \end{cases} \tag{6.18}$$

and this yields the required property of w.

By Theorem 6.1, $E(w) \neq \emptyset$ and take $W \in \mathcal{L}(\mathcal{K})$ to be an extension of w. Now we can define two isometries returning us to the space $\ell^2(\mathcal{U})$. Thus define

$$\begin{aligned} W_+ &: \ell^2(\mathcal{U}) \to \mathcal{K} \\ &W_+ h^{(n)} = W^n h^{(0)} \end{aligned} \tag{6.19}$$

and

$$\begin{aligned} W_- :& \ell^2(\mathcal{U}) \to \mathcal{K} \\ W_{-h^{(n)}} &= W^{n+1}h^{(-1)}. \end{aligned} \tag{6.20}$$

Based on (6.18), we conclude that W_+ and W_- are isometries and we define finally the operator

$$\begin{aligned} L :& \ell^2(\mathcal{U}) \to \ell^2(\mathcal{U}) \\ L &= W_-^* W_+ \end{aligned} \tag{6.21}$$

which is a contraction. Recall the shift operator on $\ell^2(\mathcal{U})$:

$$S(\cdots h_{-1}, \boxed{h_o}, h_1, \cdots) = (\cdots, h_{-2}, \boxed{h_{-1}}, h_o, h_1, \cdots) \tag{6.22}$$

and obtain

$$\begin{aligned} (LSh^{(n)}, g^{(m)}) &= (W_+Sh^{(n)}, W_{-g^{(m)}})_{\mathcal{K}} = (W_+h^{(n+1)}, W_{g^{(m)}})_{\mathcal{K}} \\ &= (W^{n+1}h^{(o)}, W^{m+1}g^{(-1)})_{\mathcal{K}} \end{aligned}$$

and

$$\begin{aligned} (SLh^{(n)}, g^{(m)}) &= (W_+h^{(n)}, W_-Sg^{(m)})_{\mathcal{K}} = (W^n h^{(0)}, W^m g^{(-1)})_{\mathcal{K}} \\ &= (W^{n+1}h^{(o)}, W^{m+1}g^{(-1)})_{\mathcal{K}} \end{aligned}$$

where we used the fact that W is a unitary operator. Consequently,

$$LS = SL. \tag{6.23}$$

The relation (6.23) implies that $L = L_f$ for a certain $f \in L^\infty(\mathcal{U})$, $\| f \| \leq 1$. Then, for $n < 0$,

$$\begin{aligned} (C_n h, g) &= (h^{(0)}, g^{(n)})_{\mathcal{H}} = (W_+h^{(0)}, W - g^{(n)})_{\mathcal{K}} \\ &= (W_-^* W_+ h^{(0)}, g^{(n)}), = (Lh^{(0)}, g^{(n)}) \\ &= (L_f h^{(0)}, g^{(n)}) = \frac{1}{2\pi} \int_0^{2\pi} e^{-int}(f(e^{it})h, g)dt \end{aligned}$$

so that f is a solution of the Nehari problem. ∎

Remark 6.5 For simplicity we supposed that $C_n \in \mathcal{L}(\mathcal{U})$, $n \leq -1$, but the same line of reasoning can be followed for $C_n \in \mathcal{L}(\mathcal{U}, \mathcal{Y})$, with $\mathcal{U}$ and $\mathcal{Y}$ different separable Hilbert spaces.

6.3 Contractive intertwining dilations

Contractive intertwining dilations provide another general abstract setting for interpolation theory. The purpose of this section is to describe some details about these. Two approaches are presented; the first one based on the idea just developed for the Nehari problem, i.e. on extending isometries and the second one, reducing the problem to a completion problem. Parametrisations for the set of all solutions are also presented. Finally, we apply the contractive intertwining dilations method to solve the Nehari, Schur and Nevanlinna–Pick interpolation problems.

Let $\mathcal{H}$ and $\mathcal{H}'$ be complex Hilbert spaces and $T \in \mathcal{L}(\mathcal{H})$, $T' \in \mathcal{L}(\mathcal{H}')$ be two contractions. By Remark 2.25, the unitary operator $U \in \mathcal{L}(\mathcal{K})$, where

$$\mathcal{K} = \cdots \bigoplus \mathcal{D}_{T^*} \bigoplus \mathcal{D}_{T^*} \bigoplus \mathcal{H} \bigoplus \mathcal{D}_T \bigoplus \cdots$$

$$U = \begin{bmatrix} & \cdot & & & & \\ & \cdot & & & & \\ & \cdot & & & & \\ \cdots & I & 0 & 0 & \cdots & \\ \cdots & 0 & D_{T^*} & T & 0 & \cdots \\ & & -T^* & D_T & 0 & \cdots \\ & & 0 & 0 & I & \cdots \\ & & & \cdot & & \\ & & & \cdot & & \\ & & & \cdot & & \end{bmatrix} \tag{6.24}$$

is the unitary dilation of the contraction T, in the sense that

$$\mathcal{K} = \bigvee_{n=-\infty}^{\infty} U^n \mathcal{H}$$

and

$$T^n = P_{\mathcal{H}}^{\mathcal{K}} U^n \mid \mathcal{H}, \quad n \geq 0.$$

Consider also the minimal isometric dilation of T, $U_+ \in \mathcal{L}(\mathcal{K}_+)$

$$\mathcal{K}_+ = \mathcal{H} \bigoplus \mathcal{D}_T \bigoplus \cdots$$

$$U_+ = \begin{bmatrix} T & 0 & \cdots \\ D_T & 0 & \cdots \\ 0 & I & \cdots \\ \cdot & & \\ \cdot & & \\ \cdot & & \end{bmatrix} \tag{6.25}$$

For the given contractions T and T', define the set

$$\mathcal{I}(T',T) = \{A \in \mathcal{L}(\mathcal{H},\mathcal{H}') \mid T'A = AT\}.$$

The problem we are faced with is the following.

'Find conditions on $A \in \mathcal{I}(T',T)$ such that the set
CID $(A) = \{A_\infty \in \mathcal{I}(U'_+,U_+) \mid \| A_\infty \| \leq 1,\ P_{\mathcal{H}'}^{\mathcal{K}'_+} A_\infty = AP_{\mathcal{H}}^{\mathcal{K}_+}\}$ is non–empty.'

Note that the condition $P_{\mathcal{H}'}^{\mathcal{K}'_+} A_\infty = AP_{\mathcal{H}}^{\mathcal{K}_+}$ means that $P_{\mathcal{H}'}^{\mathcal{K}'_+} A_\infty(I - P_{\mathcal{H}}^{\mathcal{K}_+}) = 0$ and $P_{\mathcal{H}'}^{\mathcal{K}'_+} A_\infty \mid \mathcal{H} = A$. Indeed, from the last two relations it follows that

$$P_{\mathcal{H}'}^{\mathcal{K}'_+} A_\infty = P_{\mathcal{H}'}^{\mathcal{K}'_+} A_\infty(I - P_{\mathcal{H}}^{\mathcal{K}_+}) + P_{\mathcal{H}'}^{\mathcal{K}'_+} A_\infty P_{\mathcal{H}}^{\mathcal{K}_+} = AP_{\mathcal{H}}^{\mathcal{K}_+}.$$

Conversely

$$P_{\mathcal{H}'}^{\mathcal{K}'_+} A_\infty(I - P_{\mathcal{H}}^{\mathcal{K}_+}) = AP_{\mathcal{H}}^{\mathcal{K}_+}(I - P_{\mathcal{H}}^{\mathcal{K}_+}) = 0$$

and

$$P_{\mathcal{H}'}^{\mathcal{K}'_+} A_\infty \mid \mathcal{H} = AP_{\mathcal{H}}^{\mathcal{K}_+} \mid \mathcal{H} = A.$$

The main result of this section will state that the obvious necessary condition $\| A \| \leq 1$ is also sufficient. The first proof of this is modelled after that of Theorem 6.4.

Theorem 6.6 For any contraction $A \in \mathcal{I}(T',T)$, CID$(A) \neq \emptyset$.

Proof (sketch) The positive matrix M in the proof of Theorem 6.4 is replaced by the following kernel: consider $\mathcal{H}_n = \mathcal{H}$, $n > 0$, $\mathcal{H}_n = \mathcal{H}'$, $n \leq 0$, and define for $i \leq j$ the operators

$$K_{ij} = \begin{cases} T^{j-i}, & i \geq 1 \\ T'^{(j-i)}, & j \leq 0 \\ AT^{j-i-1}, & i \leq 0, j \geq 1 \end{cases}$$

and $K_{ij} = K_{ji}^*$ for $i > j$.

It can be shown that this kernel is positive definite in a similar way to how we proved that the kernel in Remark 2.25 is positive. There is another way to prove this based on the structure of positive definite kernels on $\mathbf{Z}$, similar to the structure of Toeplitz kernels presented in Theorem 2.14. This will be described in the second volume of these notes, so we do not enter into details here. Renorming $\bigoplus_{n=-\infty}^{\infty} \mathcal{H}_n$ with this positive definite kernel, we get the Hilbert space $\hat{\mathcal{H}}$ and we consider the following two subspaces of $\hat{\mathcal{H}}$:

$$\begin{aligned} \mathcal{E} = & \text{ The subspace of } \hat{\mathcal{H}} \text{ generated by the set} \\ & \{\hat{h} = (\cdots h'_{-2}, h'_{-1}, \boxed{0}, h_1, \cdots) \mid \hat{h} \text{ has finite support}\} \end{aligned} \tag{6.27}$$

and

$$\begin{aligned} \mathcal{F} = & \text{ The subspace of } \hat{\mathcal{H}} \text{ generated by the set} \\ & \{\hat{h} = (\cdots h'_{-2}, h'_{-1}, \boxed{h'_o}, 0, h_2, \cdots) \mid h \text{ has finite support}\}. \end{aligned} \tag{6.28}$$

Then the unitary operator $w : \mathcal{E} \to \mathcal{F}$ is defined first by

$$w(\cdots h'_{-2}, h'_{-1},\ 0\ , h_1, \cdots) = (\cdots h'_{-2},\ h'_{-1}, 0, h_1, \cdots). \tag{6.29}$$

From the definition of K it follows that w is an isometry, so it extends to a unitary operator from $\mathcal{E}$ to $\mathcal{F}$. One more step is to embed $\mathcal{H}$ and $\mathcal{H}'$ into $\hat{\mathcal{H}}$. Consider

$$\begin{aligned} & i : \mathcal{H} \to \hat{\mathcal{H}} \\ & \quad ih = \text{class}(\cdots 0, \boxed{0}, h, 0, \cdots), \\ & i' : \mathcal{H}' \to \hat{\mathcal{H}} \\ & \quad i'h' = \text{class}(\cdots 0, \boxed{h'}, 0, \cdots). \end{aligned}$$

Then $\check{\mathcal{H}} = i\mathcal{H}$, $\check{\mathcal{H}}' = i'\mathcal{H}'$, $\check{T} \in \mathcal{L}(\check{\mathcal{H}})$ is defined by $\check{T}ih = i(Th)$ and $\check{T}'$ and $\check{A}$ are defined in a similar way. Finally, we get $\check{A} \in \mathcal{J}(\check{T}', \check{T})$, use Theorem 6.1 and take $W \in \mathcal{L}(\mathcal{K})$ to be a unitary extension of w. Let us describe how this W produces a solution of the present problem.

Define $\mathcal{K}_+ = \bigvee_{n=0}^{\infty} W^n \check{\mathcal{H}}$ and $\mathcal{K}'_+ = \bigvee_{n=0}^{\infty} W^n \check{\mathcal{H}}'$. Then $\mathcal{K}_+$ and $\mathcal{K}'_+$ are invariant subspaces of W and define the isometries $U_+ = W \mid \mathcal{K}_+$, $U'_+ = W \mid \mathcal{K}'_+$ on $\mathcal{K}_+$ and $\mathcal{K}'_+$

respectively. We have, for $n \geq 0$,

$$\begin{aligned}&(U_+^n(\cdots 0, \boxed{0}, h, 0, \cdots), (\cdots 0, \boxed{0}, g, 0, \cdots))_{\hat{\mathcal{H}}}\\&= ((\cdots 0, \boxed{0}, \cdots, \underset{n+1}{h}, 0, \cdots), (\cdots 0, \boxed{0}, g, 0, \cdots))_{\hat{\mathcal{H}}} = (T^n h, g) = (\check{T}^n ih, ig).\end{aligned}$$

Consequently U_+ is the minimal isometric dilation of $\check{T}$. Then,

$$\begin{aligned}&(U_+'^n(\cdots 0, \boxed{h'}, 0, \cdots), (\cdots, 0, \boxed{g'}, 0, \cdots))_{\hat{\mathcal{H}}}\\&= (W^n(\cdots 0, \boxed{h'}, 0, \cdots), (\cdots 0, \boxed{g'}, 0, \cdots))_{\hat{\mathcal{H}}}\\&= ((\cdots 0, \boxed{h'}, 0, \cdots), W^{*n}(\cdots 0, \boxed{g'}, 0, \cdots))_{\hat{\mathcal{H}}}\\&= ((\cdots 0, \boxed{h'}, 0, \cdots)\ (\cdots 0, \underset{-n}{g'}, 0, \cdots, \boxed{0}, \cdots))_{\hat{\mathcal{H}}} = (T'^n h', g')\\&= (\check{T}'^n i'h', i'g')\end{aligned}$$

Consequently U_+' is the minimal isometric dilation of $\check{T}'$. Then we return into the dilations spaces by the isometries:

$$W_+ : \mathcal{K}_+ \to \mathcal{K}$$
$$W_+ = \text{The inclusion of } \mathcal{K}_+ \text{ into } \mathcal{K}$$

and

$$W_- : \mathcal{K}_+' \to \mathcal{K}$$
$$W_- = \text{The inclusion of } \mathcal{K}_+' \text{ into } \mathcal{K}$$

and we show that

$$\check{A}_\infty = W_-^* W_+ \tag{6.30}$$

belongs to CID($\check{A}$). It is obvious that $\| \check{A}_\infty \| \leq 1$. Then, for $\check{h} \in \check{\mathcal{H}}$ and $\check{g}' \in \check{\mathcal{H}}'$, we compute

$$\begin{aligned}&(P_{\check{\mathcal{H}}'}^{\mathcal{K}_+'} \check{A}_\infty U_+^n \check{h}, \check{g}')_{\hat{\mathcal{H}}} = (\check{A}_\infty W^n \check{h}, \check{g}')_{\hat{\mathcal{H}}} = (P_{\mathcal{K}_+'}^{\mathcal{K}} W^n \check{h}, \check{g}')_{\hat{\mathcal{H}}} = (W^n \check{h}, \check{g}')_{\hat{\mathcal{H}}}\\&= (T'^n Ah, g') = (AT^n h, g') = (\check{A}\check{T}^n \check{h}, \check{g}')_{\hat{\mathcal{H}}} = (\check{A} P_{\check{\mathcal{H}}}^{\mathcal{K}_+} U_+^n \check{h}, \check{g}')_{\hat{\mathcal{H}}}\end{aligned}$$

so that

$$P_{\check{\mathcal{H}}'}^{\mathcal{K}_+'} \check{A}_\infty = \check{A} P_{\check{\mathcal{H}}}^{\mathcal{K}_+}.$$

Now let us show that $\check{A}_\infty \in \mathcal{I}(U'_+, U_+)$. For $n \geq 1$, $m \geq 0$, $\check{h} \in \check{\mathcal{H}}$, $\check{g}' \in \check{\mathcal{H}}'$, we have

$$\begin{aligned}(\check{A}_\infty U_+(U_+^m \check{h}), U_+'^n \check{g}')_{\hat{\mathcal{H}}} &= (U_+^{m+1}\check{h}, U_+'^n \check{g}')_{\hat{\mathcal{H}}} = (U_+^m \check{h}, U_+'^{(n-1)}\check{g}')_{\hat{\mathcal{H}}} \\ &= (\hat{A}_\infty U_+^m \check{h}, U_+'^{(n-1)}\check{g}')_{\hat{\mathcal{H}}} = (U'_+ \check{A}_\infty U_+^m \check{h}, U_+'^n \check{g}')_{\hat{\mathcal{H}}'},\end{aligned}$$

and for $n = 0$, $m \geq 0$,

$$\begin{aligned}(\hat{A}_\infty U_+(U_+^m \check{h}), \check{g}')_{\hat{\mathcal{H}}} &= (U_+^{m+1}\check{h}, \check{g}'_{\hat{\mathcal{H}}}) = (T'^{(m+1)}Ah, g') = (T'AT^m h, g') \\ &= (\check{A}\check{T}^m \check{h}, \check{T}'^* \check{g}')_{\hat{\mathcal{H}}} = (\check{A}P_{\mathcal{H}}^{\mathcal{K}+} U_+^m \check{h}, \check{T}'^* \check{g}')_{\hat{\mathcal{H}}} = \check{A}_\infty U_+^m \check{h}, U'^* \check{g}')_{\hat{\mathcal{H}}} \\ &= (U'_+ \check{A}_\infty U_+^m \check{h}, \check{g}')_{\hat{\mathcal{H}}}\end{aligned}$$

proving that $\check{A}_\infty \in \mathcal{I}(U'_+, U_+)$. Finally, it is clear how $\check{A}_\infty$ induces an operator $A_\infty \in \mathrm{CID}(A)$. ∎

Most of the rest of this section is devoted to the parametrisation of the set $\mathrm{CID}(A)$.

Theorem 6.7 There exists a one–to–one correspondence between $\mathrm{CID}(A)$ and $E(w)$, where w is given by (6.29).

Proof We continue with the same notation as in the proof of Theorem 6.6. Let W be the considered extension of w; then for $n \in \mathbf{Z}$ the compressions $P_{\check{\mathcal{H}}} W^n \mid \check{\mathcal{H}}$ are determined by $\check{T}$, because U_+ is the minimal isometric dilation of $\check{T}$, and similarly the compressions $P_{\check{\mathcal{H}}'} W^n \mid \check{\mathcal{H}}'$ are determined by $\check{T}'$. Then, for $m, p \geq 0$, $\check{h} \in \check{\mathcal{H}}$, $\check{g}' \in \check{\mathcal{H}}'$,

$$(\check{A}_\infty U_+^m \check{h}, U_+'^p \check{g}')_{\hat{\mathcal{H}}} = (U_+^m \check{h}, U_+'^p \check{g}')_{\hat{\mathcal{H}}} = (W^{m-p}\check{h}, \check{g}')_{\hat{\mathcal{H}}}.$$

Consequently, the compressions $P_{\check{\mathcal{H}}'} W^n \mid \check{\mathcal{H}}$ and $P_{\check{\mathcal{H}}} W^n \mid \check{\mathcal{H}}'$ are determined by $\check{A}_\infty$. But now, by the minimality of W and by $\check{\mathcal{H}} \bigvee \check{\mathcal{H}}' \subset \hat{\mathcal{H}}$ it follows that $\bigvee_{n=-\infty}^{\infty} W^n(\check{\mathcal{H}} \bigvee \check{\mathcal{H}}') \subset \mathcal{K}$. On the other hand, by the definition of w, $\mathcal{K} \subset \bigvee_{n=0}^{\infty} w^n \check{\mathcal{H}} \bigvee \bigvee_{n=0}^{\infty} w^{*n} \check{\mathcal{H}}'$ so that

$$\mathcal{K} = \bigvee_{n=-\infty}^{\infty} W^n(\check{\mathcal{H}} \bigvee \check{\mathcal{H}}'). \tag{6.31)a$$

Consequently, if $W \in \mathcal{L}(\mathcal{K})$ and $W' \in \mathcal{L}(\mathcal{K}')$ produce the same $\check{A}_\infty$, then the operator

$$\omega(\sum W^n h_n) = \sum W'^n h_n, \quad h_n \in \check{\mathcal{H}} \bigvee \check{\mathcal{H}}'$$

extends to a unitary operator from $\mathcal{K}$ onto $\mathcal{K}'$ with $\omega W = W'\omega$. Furthermore since W and W' extend w, $\omega \mid \hat{\mathcal{H}} = I$ and thus, W and W' determine the same element in $E(w)$.

Consider now $\check{A}_\infty \in \mathrm{CID}(\check{A})$ and construct the kernel on $\mathbf{Z}$, by

$$\hat{K}_n = \begin{bmatrix} U_+'^n & \check{A}_\infty U_+^n \\ \check{A}_\infty U_+'^n & U_+^n \end{bmatrix} = \begin{bmatrix} I & \check{A}_\infty \\ \check{A}_\infty^* & I \end{bmatrix} \begin{bmatrix} U_+'^n & 0 \\ 0 & U_+^n \end{bmatrix}.$$

But, by the properties of $\check{A}_\infty$,

$$\begin{bmatrix} U_+'^* & 0 \\ 0 & U_+^* \end{bmatrix} \begin{bmatrix} I & \check{A}_\infty \\ \check{A}_\infty^* & I \end{bmatrix} \begin{bmatrix} U_+' & 0 \\ 0 & U_+ \end{bmatrix} = \begin{bmatrix} I & \check{A}_\infty \\ \check{A}_\infty^* & I \end{bmatrix}.$$

It follows by Proposition A.2 that there exists a contraction Γ such that

$$\begin{bmatrix} I & \check{A}_\infty \\ \check{A}_\infty^* & I \end{bmatrix}^{\frac{1}{2}} \begin{bmatrix} U_+' & 0 \\ 0 & U_+ \end{bmatrix} = \Gamma \begin{bmatrix} I & \check{A}_\infty \\ \check{A}_\infty^* & I \end{bmatrix}^{\frac{1}{2}}$$

and then

$$\hat{K}_n = \begin{bmatrix} I & \check{A}_\infty \\ \check{A}_\infty^* & I \end{bmatrix}^{\frac{1}{2}} \Gamma^n \begin{bmatrix} I & \check{A}_\infty \\ \check{A}_\infty^* & I \end{bmatrix}^{\frac{1}{2}}.$$

According to Theorem 2.14, the kernel $\hat{K}_n$ is a positive definite Toeplitz kernel. Moreover, Proposition 2.20 yields a Hilbert space $\hat{\mathcal{K}}$ and a unitary operator $\hat{W}$ on $\hat{\mathcal{K}}$ such that

$$\hat{K}_n = P^{\hat{\mathcal{K}}}_{\mathcal{K}'_+ \oplus \mathcal{K}_+} \hat{W}^n \mid \mathcal{K}'_+ \bigoplus \mathcal{K}_+, \quad n \geq 0.$$

Now, we have only to remark that, since the kernel defined by (6.26) appears as a compression of $\hat{K}$, $\hat{W}$ is a unitary extension of w and the application

$$E(w) \ni W \to \check{A}_\infty \in \mathrm{CID}(\check{A})$$

used in the proof of Theorem 6.6 is one–to–one. As the correspondence between $\mathrm{CID}(\check{A})$ and $\mathrm{CID}(A)$ is one–to–one, we get the required result. ∎

Corollary 6.8 There exists a one–to–one correspondence between $\mathrm{CID}(A)$ and the sequences of contractions$\{G_n\}_{n=1}^\infty$, $G_1 \in \mathcal{L}(\hat{\mathcal{H}} \ominus \mathcal{E}, \hat{\mathcal{H}} \ominus \mathcal{F})$, $G_n \in \mathcal{L}(\mathcal{D}_{G_{n-1}}, \mathcal{D}_{G^*_{n-1}})$, where $\mathcal{E}, \mathcal{F}$ and $\hat{\mathcal{H}}$ are the spaces defined in the proof of Theorem 6.6

We will obtain another proof of Theorem 6.6. For this purpose, define

$$\mathcal{K}_n = \mathcal{H} \bigoplus \bigoplus_{k=1}^{n-1} \mathcal{D}_T, \quad P_n = P^{\mathcal{K}_+}_{\mathcal{K}_n} \tag{6.32}$$

$$T_n = P_n U_+ \mid \mathcal{K}_n \tag{6.33}$$

and similar definitions hold for T'. We immediately get

$$\mathcal{K} = \bigvee_{n=0}^{\infty} \mathcal{K}_n \tag{6.34}$$

$$U_+ = \text{s} - \lim_{n\to\infty} T_n P_n. \tag{6.35}$$

Consider the following sets:

$$\text{n} - \text{PCID}(A) = \{A_n \in \mathcal{I}(T'_n, T_n \mid \| A_n \| \leq 1, P'_o A_n = AP_o \mid \mathcal{K}_n\}$$

and

$$\text{PCID}(A) = \{\{A_n\}_{n=0}^{\infty} \mid A_n \in \text{n} - \text{PCID}(A), P'_n A_{n+1} = A_n P_n \mid \mathcal{K}_{n+1}\}.$$

Proposition 6.9 The formula

$$\{A_n\}_{n=0}^{\infty} \to A_\infty = \text{s} - \lim_{n\to\infty} A_n P_n$$

establishes a one–to–one correspondence between PCID(A) and CID(A).

Proof To begin with, let us note the matrix structure of an element A_n of a sequence $\{A_n\}_{n=0}^{\infty}$ in PCID(A). Taking into account the remark just after the definition of CID(A),

$$A_n = \begin{bmatrix} A & 0 \\ * & * \end{bmatrix}.$$

Furthermore, taking into account the commutativity property of A_n and the matrix structure of the operators T_n and T'_n, it follows that A_n has the following lower triangular structure:

$$A_n = \begin{bmatrix} A & 0 & \cdots & 0 \\ X_{11} & X_{12} & 0, \cdots & 0 \\ \cdot & & & \\ \cdot & & & \\ \cdot & & & \\ X_{n1} & X_{n2} & \cdots & X_{nn} \end{bmatrix} : \mathcal{K}_n \to \mathcal{K}_n \tag{6.36}$$

where

$$X_{11}T + X_{12}D_T = D_{T'}A \tag{6.37}$$

$$X_{k1}T + X_{k2}D_T = X_{k-1,1} \tag{6.38}$$

$$X_{kp} = X_{k-1,p-1} \quad \text{for } n \geq k \geq 2,\ 3 \leq p \leq k+1. \tag{6.39}$$

From this lower triangular structure it follows that

$$\text{s} - \lim_{n \to \infty} A_n P_n = A_\infty$$

exists and $\| A_\infty \| \leq 1$. Moreover, from $T'_n A_n = A_n T_n$ it follows that $T'_n P'_n A_n P_n = A_n P_n T_n P_n$ and considering the strong limit when $n \to \infty$, we get $A_\infty \in \mathcal{I}(U'_+, U_+)$. In the same way, it follows that $P'_o A_\infty = AP_o$. Conversely, consider $A_\infty \in \text{CID}(A)$. Then $\{P'_n A_\infty \mid \mathcal{K}_n\}_{n=0}^\infty \in \text{PCID}(A)$ and $A_\infty = \text{s} - \lim_{n\to\infty} P'_n A_\infty P_n$. ∎

Remark 6.10 Another proof of Theorem 6.6 can be immediately obtained. The main idea is to turn the necessary condition $A_1 \in 1 - \text{PCID}(A)$ into a completion problem. For this purpose, by (6.37) and (6.38) we have

$$A_1 = \begin{bmatrix} A & 0 \\ X_{11} & X_{12} \end{bmatrix}$$

and $X_{11}T + X_{12}D_T = D_{T'}A$. Multiplying A_1 on the right with $J(T)$, we get the operator

$$B_1 = \begin{bmatrix} AT & AD_{T^*} \\ D'_T A & X_{11}D_{T^*} - X_{12}T^* \end{bmatrix}.$$

Once we have determined the operator $S_1 = X_{11}D_{T^*} - X_{12}T^* : \mathcal{D}_{T^*} \to \mathcal{D}_{T'}$ such that B_1 is a contraction, X_{11} and X_{12} are determined to satisfy the required conditions. But now,

$$(AT, AD_{T^*})(AT, AD_{T^*})^* = ATT^*A^* + AA^* - ATT^*A^* = AA^* \leq I$$

and

$$(T^*A^*, A^*D_{T'})(T^*A^*, A^*D_{T'})^* = A^*T'^*T'A + A^*A - A^*T'^*T'A = A^*A \leq I$$

and the existence of an S_1 making B_1 a contraction is assured by Theorem A.7.

To finish the proof of Theorem 6.6, we can choose $A_n \in 1 - \text{PCID}(A_{n-1})$ for $n \geq 1$. Then $\{A_n\}_{n=0}^\infty \in \text{PCID}(A)$ and by Proposition 6.9, $\text{CID}(A) \neq \emptyset$.

We continue to develop Remark 6.10. Define

$$S_n = X_{n1}D_{T^*} - X_{n2}T^* : \mathcal{D}_{T^*} \to \mathcal{D}_{T'}. \tag{6.40}$$

Then $\{A_n\}_{n=0}^{\infty} \in \mathrm{PCID}(A)$ if and only if the operators defined by the matrices

$$B_n = \begin{bmatrix} AT^n & AT^{n-1}D_{T^*} & \cdots & & AD_{T^*} \\ D_{T'}AT^{n-1} & D_TAT^{n-2}D_{T^*} & \cdots & & S_1 \\ & \cdots & \cdots & S_1 & S_2 \\ \cdot & & & & \\ \cdot & & & & \\ \cdot & & & & \\ D_{T'}A & S_1 & S_2\cdots & S_{n-1} & S_n \end{bmatrix} \tag{6.41}$$

are contractions. Indeed, it follows from the definitions that

$$(X_{n-1,1}, S_n) = (X_{n1}, X_{n2})J(T)$$

and multiplying A_n sufficiently many times on the right with operators $J(T)$ augmented with identities on convenient spaces, we obtain exactly B_n. In this way the study of $\mathrm{CID}(A)$ is reduced to the analysis of the operators B_n, and now this closely resembles the analysis in Theorem 2.1. We begin with B_1. We denote $C_o = AT$ and we search for two operators $X \in \mathcal{L}(\mathcal{D}_{C_o}, \mathcal{D}_{T'})$ and $Y \in \mathcal{L}(\mathcal{D}_{T^*}, \mathcal{D}_{C_o^*})$ such that

$$B_1 = \begin{bmatrix} C_o & D_{C_o^*}Y \\ XD_{C_o} & S_1 \end{bmatrix}.$$

In this form we would have only to use Theorem A.7 in order to obtain the structure of S_1. For this purpose, we consider the unitary operators

$$\begin{aligned} &\omega^{A^*} : \mathcal{D}_{C_o^*} \to \mathcal{F}^{A^*} \\ &\qquad \omega^{A^*} D_{C_o^*} = F^{A^*} = (D_{T^*}A^*, D_{A^*})^{\mathrm{t}} \end{aligned} \tag{6.42}$$

where $\mathcal{F}^{A^*} = \overline{\mathcal{R}(F^{A^*})} \subset \mathcal{D}_{T^*} \bigoplus \mathcal{D}_{A^*}$, and

$$\begin{aligned} &\omega^{A} : \mathcal{D}_{C_o} \to \mathcal{F}^{A} \\ &\qquad \omega^{A} D_{C_o} = F^{A} = (D_{T'}A, D_A)^{\mathrm{t}} \end{aligned} \tag{6.43}$$

where $\mathcal{F}^A = \overline{\mathcal{R}(F^A)} \subset \mathcal{D}_{T'} \bigoplus \mathcal{D}_A$.

We obtain

$$AD_{T^*} = (F^{A^*})^*(I,0)^{\mathrm{t}} = D_{C_o^*}(\omega^{A^*})^* p^{A^*}(I,0)^{\mathrm{t}}$$

where $p^{A^*} = P_{\mathcal{F}^{A^*}}^{\mathcal{D}_{T^*} \bigoplus \mathcal{D}_{A^*}}$. Consequently, we can take

$$Y = (\omega^{A^*})^* p^{A^*}(I,0)^{\mathrm{t}} : \mathcal{D}_{T^*} \to \mathcal{D}_{C_o^*}. \tag{6.44}$$

Similarly,

$$D_{T'}A = (I,0)F^A = (I,0)p^A\omega^A D_{C_o}$$

where $p^A = P_{\mathcal{F}^A}^{\mathcal{D}_{T'} \bigoplus \mathcal{D}_A}$ and we can take

$$X = (I,0)p^A\omega^A : \mathcal{D}_{C_o} \to \mathcal{D}_{T'}. \tag{6.45}$$

For determining the structure of B_n we need some more notation. Representing B_n in the form

$$B_n = \begin{bmatrix} C_{n-1} & r_{n-1} \\ s_{n-1} & S_n \end{bmatrix} \tag{6.46}$$

we have defined the block matrices $C_n, r_n, s_n \geq 0$. Moreover, we define

$$V = (T, D_{T^*}) \tag{6.47}$$

and

$$V = (T', D_{T'})^{\mathrm{t}}. \tag{6.48}$$

We can now prove a result corresponding to the first part of the proof of Theorem 2.1.

Proposition 6.11 There exist a family of contractions $\{G_n\}_{n=1}^{\infty}$, $G_1 \in \mathcal{L}(\mathcal{D}_Y, \mathcal{D}_{X^*})$, $G_n \in \mathcal{L}(\mathcal{D}_{G_{n-1}}, \mathcal{D}_{G_{n-1}^*})$ and two families of unitary operators

$$\Omega_n : \mathcal{D}_{C_n} \to \mathcal{D}_{C_o} \bigoplus \mathcal{D}_Y \bigoplus \mathcal{D}_{G_1} \bigoplus \cdots \bigoplus \mathcal{D}_{G_{n-1}}$$
$$\tilde{\Omega} : \mathcal{D}_{C_n^*} \to \mathcal{D}_{C_o^*} \bigoplus \mathcal{D}_{X^*} \bigoplus \mathcal{D}_{G_1^*} \bigoplus \cdots \bigoplus \mathcal{D}_{G_{n-1}^*}$$

such that

$$S_1 = -XC_o^*Y + D_{X^*}G_1D_Y$$

and, for $n \geq 2$

$$S_n = -L(X, G_1, \cdots, G_{n-1})\Omega_{n-1}C_{n-1}^*\tilde{\Omega}_{n-1}^*\tilde{L}(Y, G_1, \cdots, G_{n-1})$$
$$+ D_{X^*}D_{G_1^*}\cdots D_{G_{n-1}^*}G_nD_{G_{n-1}}\cdots D_{G_1}D_Y.$$

Moreover, the correspondence between $\{S_n\}_{n=1}^{\infty}$ making B_n, $n \geq 1$, contractions and $\{G_n\}_{n=1}^{\infty}$ is one to one.

Proof For $n = 0$, we simply have $\Omega_o = I_{\mathcal{D}_{C_0}}$, $\tilde{\Omega}_o = I_{\mathcal{D}_{C_o^*}}$ and for $n = 1$ we use Theorem A.7. Furthermore, we prove by induction the following statements: for $n \geq 2$,

$$\left\{\begin{array}{l}\text{There exist the unitary operators}\\ \Omega_{n-1} : \mathcal{D}_{C_{n-1}} \to \mathcal{D}_{C_0} \bigoplus \mathcal{D}_Y \bigoplus \mathcal{D}_{G_1} \bigoplus \cdots \bigoplus \mathcal{D}_{G_{n-2}}\\ \text{and the relations } s_{n-1} = L(X, G_1, \cdots, G_{n-1})\Omega_{n-1} D_{C_{n-1}} \text{ hold.}\end{array}\right. \qquad (6.49)_n$$

From now on we denote $L(X, G_1, \cdots, G_{n-1})$ by L_{n-1}):

$$\left\{\begin{array}{l}\text{There exist the unitary operators}\\ \tilde{\Omega}_{n-1} : \mathcal{D}_{C^*_{n-1}} \to \mathcal{D}_{C_o^*} \bigoplus \mathcal{D}_{X^*} \bigoplus \mathcal{D}_{G_1^*} \bigoplus \cdots \bigoplus \mathcal{D}_{G^*_{n-2}}\\ \text{and the relations } r_{n-1} = D_{C^*_{n-1}} \tilde{\Omega}^*_{n-1} \tilde{L}(Y, G_1, \cdots, G_{n-1}) \text{ hold}\end{array}\right. \qquad (6.50)_n$$

(and denote $\tilde{L}(Y, G_1, \cdots, G_{n-1})$ by $\tilde{L}_{n-1}$):

$$\left\{\begin{array}{l}\text{There exist the uniquely determined contractions}\\ G : \mathcal{D}_{G_{n-1}} \to \mathcal{D}_{G^*_{n-1}} \text{ such that}\\ S_n = -L_{n-1}\Omega_{n-1}C^*_{n-1}\tilde{\Omega}^*_{n-1}\tilde{L}_{n-1}\\ \qquad +D_{X^*}D_{G_1^*}\cdots D_{G^*_{n-1}} G_n D_{G_{n-1}} \cdots D_{G_1} D_Y.\end{array}\right. \qquad (6.51)_n$$

The case $n = 2$ is immediately verified and the general case is as follows. By direct computation using the definitions, we have

$$C_n = (C_{n-1}, s_{n-1})^{t}(V \bigoplus I_{n-1}) = (V' \bigoplus I_{n-1})(C_{n-1}, r_{n-1}) \qquad (6.52)$$

and, by the results in the Appendix, we obtain the following diagram defining the operator Ω_n:

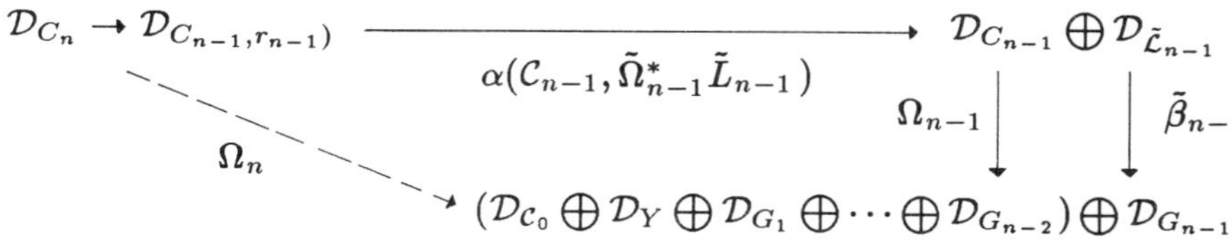

$$\begin{array}{ccc} \mathcal{D}_{C_n} \to \mathcal{D}_{C_{n-1}, r_{n-1})} & \xrightarrow{\alpha(C_{n-1}, \tilde{\Omega}^*_{n-1}\tilde{L}_{n-1})} & \mathcal{D}_{C_{n-1}} \bigoplus \mathcal{D}_{\tilde{\mathcal{L}}_{n-1}} \\ & \searrow \Omega_n & \Omega_{n-1}\downarrow \qquad \downarrow \tilde{\beta}_{n-} \\ & & (\mathcal{D}_{C_0} \bigoplus \mathcal{D}_Y \bigoplus \mathcal{D}_{G_1} \bigoplus \cdots \bigoplus \mathcal{D}_{G_{n-2}}) \bigoplus \mathcal{D}_{G_{n-1}} \end{array}$$

and we get the formula

$$\Omega_n D_{C_n} = \begin{bmatrix} \Omega_{n-1} D_{C_{n-1}} & -\Omega_{n-1} C^*_{n-1} \tilde{\Omega}^*_{n-1} \tilde{L}_{n-1} \\ 0 & D_{G_{n-1}} \cdots D_{G_1} D_Y \end{bmatrix}. \qquad (6.53)$$

But now

$$\begin{aligned}s_n = (s_{n-1}, S_n) &= (L_{n-1}\Omega_{n-1}D_{C_{n-1}}, -L_{n-1}\Omega_{n-1}C^*_{n-1}\tilde{\Omega}^*_{n-1}\tilde{L}_{n-1}\\ &\quad + D_{X^*}D_{G_1^*}\cdots D_{G^*_{n-1}}G_nD_{G_{n-1}}\cdots D_{G_1}D_Y)\\ &= L_n\Omega_nD_{C_n}.\end{aligned}$$

The relations $(6.50)_n$ follow in a similar way, and the recurrence formulas for $\tilde{\Omega}_nD_{C_n^*}$ are,

$$\tilde{\Omega}_nD_{C_n^*} = \begin{bmatrix}\tilde{\Omega}_{n-1}D_{C^*_{n-1}} & -\tilde{\Omega}_{n-1}C_{n-1}\Omega^*_{n-1}L^*_{n-1}\\ 0 & D_{G^*_{n-1}}\cdots D_{G_1^*}D_{X^*}\end{bmatrix} \tag{6.54}$$

Finally, $(6.51)_n$ follows from $(6.49)_n$, $(6.50)_n$ and Theorem A.7. ■

Consider now the operators

$$\begin{gathered}Q_n : \mathcal{D}_{C_o^*}\bigoplus\mathcal{D}_{X^*}\bigoplus\bigoplus_{k=1}^{n-1}\mathcal{D}_{G_k^*}\to\mathcal{D}_{C_o}\bigoplus\mathcal{D}_Y\bigoplus\bigoplus_{k=1}^{n-1}\mathcal{D}_{G_k}\\ Q_n = -\Omega_nC_n\tilde{\Omega}^*_n.\end{gathered} \tag{6.55}$$

We obviously have $Q_o = -C_o^*$ and let us determine Q_1. For this purpose, we compute

$$\begin{aligned}\Omega_1C_1^*\tilde{\Omega}_1^*\begin{bmatrix}D_{C_o^*} & -C_oX^*\\ 0 & D_{X^*}\end{bmatrix} &= \Omega_1C_1^*D_{C_1^*} = \Omega_1D_{C_1}C_1^* = \begin{bmatrix}D_{C_o} & -C_o^*Y\\ 0 & D_Y\end{bmatrix}\\ &= \alpha(C_o,Y)D_{L(C_o,Y)}V^*\tilde{L}(C_o,X)^*.\end{aligned} \tag{6.56}$$

But now we remark that

$$\begin{aligned}D^2_{L(C_o,Y)}V^* &= V^* - (T^*A^*, D_{T^*}A^*)^{\mathrm{t}}(AT, AD_{T^*})(T^*, D_{T^*})^{\mathrm{t}}\\ &= V^*D_A^2 = V^*D^2_{\tilde{L}(C_o,X)}\end{aligned}$$

and it follows by a standard argument that

$$D_{L(C_o,Y)}V^* = V^*D_{\tilde{L}(C_o,X)}. \tag{6.57}$$

This implies that $V^*\mathcal{D}_{\tilde{L}(C_o,X)}\subset\mathcal{D}_{L(C_o,Y)}$ and we can continue in (6.56) by

$$\begin{aligned}\alpha(C_o,Y)D_{L(C_o,Y)}V^*\tilde{L}(C_o,X)^* &= \alpha(C_o,Y)V^*D_{\tilde{L}(C_o,X)}\tilde{L}(C_o,X)^*\\ &= \alpha(C_o,Y)V^*\tilde{L}(C_o,X)^*D_{\tilde{L}(C_o,X))^*}\\ &= \alpha(C_o,Y)V^*\tilde{L}(C_o,X)^*\tilde{\alpha}^*(C_o,X)\tilde{\alpha}(C_o,X)D_{\tilde{L}(C_o,X)^*}.\end{aligned}$$

Using Lemma A.8, this means that

$$\begin{aligned}\Omega_1 C_1^* \tilde{\Omega}_1^* &= \alpha(C_o, Y) V^* \tilde{L}(C_o, X)^* \tilde{\alpha}(C_o, X)^* \\ &= \alpha(C_o, Y) V^* \tilde{\beta}^*(C_o, X)(D_X C_o^*, X^*).\end{aligned}$$

Finally we get

$$Q_1 = V_o \begin{bmatrix} Q_o & 0 \\ 0 & I \end{bmatrix} \tag{6.58}$$

where

$$\begin{aligned}&V_o : \mathcal{D}_{C_o} \bigoplus \mathcal{D}_{X^*} \to \mathcal{D}_{C_o} \bigoplus \mathcal{D}_Y \\ &V_o = \alpha(C_o, Y) V^* \tilde{\beta}^*(C_o, X)(D_X, -X^*).\end{aligned} \tag{6.59}$$

With these preliminaries we can obtain the main result concerning the parametrisation of CID(A).

Theorem 6.12 There exists a one–to–one correspondence between CID(A) and the sequences of contractions $\{G_n\}_{n=1}^{\infty}$, $G_1 \in \mathcal{L}(\mathcal{D}_Y, \mathcal{D}_{X^*})$, $G_n \in \mathcal{L}(\mathcal{D}_{G_{n-1}}, \mathcal{D}_{G^*_{n-1}})$ $n \geq 2$, such that

$$S_n = L_{n-1} Q_{n-1} \tilde{L}_{n-1} + D_{X^*} D_{G_1^*} \cdots D_{G^*_{n-1}} G_n D_{G_{n-1}} \cdots D_{G_1} D_Y$$

where

$$Q_o = -C_o^*, \quad Q_1 = V_o(Q_o \bigoplus I)$$

and for $n \geq 2$,

$$Q_n = (V_o \bigoplus I_{n-1})(I \bigoplus V_{n-1})(Q_{n-1} \bigoplus I),$$

where V_o is the operator defined by (6.59) and $V_n = V(\{G_k\}_{k-1}^{n})$ are the operators defined by (A.2).

Proof We have by the definitions, (6.52) and Lemma A.8 that

$$\begin{aligned}Q_n &= -\Omega_n(V^* \bigoplus I_{n-1})\tilde{\beta}^*(C_{n-1}, L_{n-1}\Omega_{n-1}) \\ &\quad (\tilde{\beta}(C_{n-1}, L_{n-1}\Omega_{n-1})\tilde{L}(C_{n-1}, L_{n-1}\Omega_{n-1})^* \tilde{\alpha}^*(C_{n-1}, L_{n-1}\Omega_{n-1})) \\ &\quad (\tilde{\Omega}^*_{n-1} \bigoplus \beta^*_{n-1}) \\ &= -\Omega_n(V^* \bigoplus I_{n-1})\tilde{\beta}^*(C_{n-1}, L_{n-1}\Omega_{n-1}) \\ &\quad (\Omega^*_{n-1} D_{L_{n-1}} \Omega_{n-1} C^*_{n-1}, \Omega^*_{n-1} L^*_{n-1})(\tilde{\Omega}^*_{n-1} \bigoplus \beta^*_{n-1}) \\ &= \Omega_n(V^* \bigoplus I_{n-1})\tilde{\beta}^*(C_{n-1}, L_{n-1}\Omega_{n-1})\tilde{\Omega}^*_{n-1}(D_{L_{n-1}}, L^*_{n-1}\beta^*_{n-1})(Q_{n-1} \bigoplus I)\end{aligned}$$

so that we have only to prove the identity

$$\begin{aligned} V(n) &= \Omega_n(V^* \bigoplus I_{n-1})\tilde{\beta}^*(C_{n-1}, L_{n-1}\Omega_{n-1})\tilde{\Omega}^*_{n-1}(D_{L_{n-1}}, L^*_{n-1}\beta^*_{n-1}) \\ &= (V_o \bigoplus I_{n-1})(I \bigoplus V_{n-1}). \end{aligned} \tag{6.60}$$

Note first that similar to (6.57), we have

$$D_{L(C_{n-1},\tilde{\Omega}^*_{n-1}\tilde{L}_{n-1})}(V^* \bigoplus I_{n-1}) = (V^* \bigoplus I_{n-1})D_{\tilde{L}(C_{n-1},L_{n-1}\Omega_{n-1})} \tag{6.61}$$

and now, on the one hand,

$$\begin{aligned} &\Omega_n(V^* \bigoplus I_{n-1})\tilde{\beta}^*(C_{n-1}, L_{n-1}\Omega_{n-1})\tilde{\Omega}^*_{n-1}D_{L_{n-1}}\Omega_{n-1}D_{C_{n-1}} \\ &= \Omega_n(V^* \bigoplus I_{n-1})D_{\tilde{L}(C_{n-1},L_{n-1}\Omega_{n-1})} \\ &= \Omega_n D_{L(C_{n-1},\tilde{\Omega}^*_{n-1}\tilde{L}_{n-1})}(V^* \bigoplus I_{n-1}) \\ &= (V(n-1) \bigoplus I)\begin{bmatrix} I_{n-2} & 0 \\ 0 & G_{n-1} \\ 0 & D_{G_{n-1}} \end{bmatrix}\Omega_{n-1}D_{C_{n-1}}, \end{aligned}$$

and on the other hand,

$$\begin{aligned} &-\Omega_n(V^* \bigoplus I_{n-1})\tilde{\beta}^*(C_{n-1}, L_{n-1}\Omega_{n-1})\tilde{\Omega}^*_{n-1}L^*_{n-1}\beta^*_{n-1}D_{G^*_{n-1}}\cdots D_{G^*_1}D_{X^*} \\ &= -\Omega_n(V^* \bigoplus I_{n-1})\tilde{\beta}^*(C_{n-1}, L_{n-1}\Omega_{n-1})\tilde{\Omega}^*_{n-1}D_{L_{n-1}}L^*_{n-1} \\ &= -(V(n-1) \bigoplus I)\begin{bmatrix} L^*_{n-2} \\ G_{n-1}G^*_{n-1}D_{G^*_{n-2}}\cdots D_{G^*_1}D_{X^*} \\ G^*_{n-1}D_{G^*_{n-1}}D_{G^*_{n-2}}\cdots D_{G^*_1}D_{X^*} \end{bmatrix}. \end{aligned}$$

But

$$\begin{aligned} &-(D_{L_{n-2}}, -L^*_{n-2}\beta^*_{n-2})(L^*_{n-2}, G_{n-1}G^*_{n-1}D_{G^*_{n-2}}\cdots D_{G^*_1}D_{X^*})^{\mathrm{t}} \\ &= -L^*_{n-2}\beta^*_{n-2}D^2_{G^*_{n-1}}D_{G^*_{n-2}}\cdots D_{G^*_1}D_{X^*}. \end{aligned}$$

Consequently,

$$\begin{aligned} -\Omega_n(V^* \bigoplus I_{n-1})\tilde{\beta}^*(C_{n-1},L_{n-1}\Omega_{n-1})\tilde{\Omega}^*_{n-1}\tilde{L}^*_{n-1}\beta^*_{n-1} \\ = (V(n-1) \bigoplus I)(0, D_{G^*_{n-1}}, -G^*_{n-1})^{\mathrm{t}} \end{aligned}$$

and finally,

$$V(n) = (V(n-1) \bigoplus I)(I \bigoplus J(G_{n-1}))$$

which, by the use of Proposition A.9, ends the proof of (6.60). Using (6.60) and Proposition 6.11 we complete the proof of the required formula. ∎

We have seen in Chapter 2 that there exists a connection between formulas of the type in Theorem 6.12 and cascade transforms. For the present case, consider the function

$$S(z) = S_1 + zS_2 + z^2 S_3 + \cdots$$

where S_n are defined by (6.40) and this is analytic since S_n are contractions. By a comparison with Theorem 2.1, we get the following result.

Theorem 6.13 $S(z) = a(z) + b(z)s(z)(I - d(z)s(z))^{-1}c(z), \quad z \in \mathbf{D}$, where

$$\begin{aligned}
a(z) &= X(I - zA)^{-1}Q_oY : \mathcal{D}_{T^*} \to \mathcal{D}_{T'} \\
b(z) &= D_{X^*} + zX(I - zA)^{-1}B : \mathcal{D}_{X^*} \to \mathcal{D}_{T'} \\
c(z) &= D_Y + zC(I - zA)^{-1}Q_oY : \mathcal{D}_{T^*} \to \mathcal{D}_Y \\
d(z) &= z(D + zC(I - zA)^{-1}B) : \mathcal{D}_{X^*} \to \mathcal{D}_Y \\
V_o &= \begin{bmatrix} A & B \\ C & D \end{bmatrix} : \begin{matrix} \mathcal{D}_{C_o} \\ \bigoplus \\ \mathcal{D}_{X^*} \end{matrix} \to \begin{matrix} \mathcal{D}_{C_o} \\ \bigoplus \\ \mathcal{D}_Y \end{matrix}
\end{aligned}$$

and $s \in \mathcal{S}(\mathcal{D}_Y, \mathcal{D}_{X^*})$ has the Schur parameters $\{G_n\}_{n-1}^{\infty}$.

Remark 6.14 There exists the possibility of identifying the spaces $\mathcal{D}_{X^*}$ and $\mathcal{D}_Y$ in terms of the given data A, T and T'. Thus we have

$$I_{\mathcal{D}_{T'}} - XX^* = (I, 0)(I - p^A)(I, 0)^t$$

and let us consider the operator

$$\begin{aligned}
\gamma_{X^*} &: \mathcal{D}_{X^*} \to \overline{(I - p^A)\mathcal{D}_{T'}} \\
\gamma_{X^*} D_{X^*} &= (I - p^A)(D_T, 0)^t
\end{aligned} \tag{6.62}$$

which is unitary (here $\mathcal{D}_{T'}$ is viewed as $\mathcal{D}_{T'} \bigoplus 0$ in $\mathcal{D}_{T'} \bigoplus \mathcal{D}_A$).

But now, we can prove that

$$\overline{(I - p^A)\mathcal{D}_{T'}} = \mathcal{R}^A = (\mathcal{D}_{T'} \oplus \mathcal{D}_A) \ominus \mathcal{F}^A. \tag{6.63}$$

For this, let $r = d' \oplus d \in \mathcal{R}^A$ and $r \perp \overline{(I - p^A)\mathcal{D}_{T'}}$. Then $(r, (I - p^A)(d'', 0)) = 0$ for any $d'' \in \mathcal{D}_{T'}$. Furthermore,

$$0 = (r, (I - p^A)(d'', 0)) = ((I - p^A)r, (d'', 0)) = (r, (d'', 0)) = (d', d''),$$

and consequently, $d' = 0$. But $r \perp \mathcal{F}^A$ means $D_A d + A^* D_{T'} d' = 0$ so $D_A d = 0$ and, since $d \in \mathcal{D}_A$, it follows that $d = 0$ and (6.63) is proved. In a similar way we have the unitary operator

$$\begin{aligned} &\gamma_Y : \mathcal{D}_Y \to \overline{(I - p^{A^*})\mathcal{D}_{T^*}} \\ &\gamma_Y D_Y = (I - p^{A^*})(D_{T^*}, 0)^{\mathrm{t}} \end{aligned} \tag{6.64}$$

and

$$\overline{(I - p^{A^*})\mathcal{D}_{T^*}} = \mathcal{R}^{A^*} = (\mathcal{D}_{T^*} \oplus \mathcal{D}_{A^*}) \ominus \mathcal{F}^{A^*}. \tag{6.65}$$

These considerations also give a criterion for uniqueness in terms of given data.Thus, by Corollary 6.8, CID(A) contains a unique element if and only if either $\mathcal{R}^A = 0$ or $\mathcal{R}^{A^*} = 0$.

We now show that the problems of Nehari, Schur and Nevanlinna–Pick can be viewed as particular cases of the lifting of commutants. First, we take into account the Schur problem in its operatorial version in Section 2.2. Let $C_k \in \mathcal{L}(\mathcal{H}_1, \mathcal{H}_2)$, $k = 0, \cdots, n$ be given operators and consider the Toeplitz block matrix

$$T_n = \begin{bmatrix} C_o & 0 & \cdots & 0 \\ C_1 & C_1 & \cdots & 0 \\ \cdot & & & \\ \cdot & & & \\ \cdot & & & \\ C_n & C_{n-1} & \cdots & C_o \end{bmatrix}$$

Moreover, consider the truncated translation

$$S_n(\mathcal{H}_i) = \begin{bmatrix} 0 & 0 & \cdots & & 0 \\ I & 0 & \cdots & & 0 \\ 0 & I & \cdots & & \\ \cdot & & & & \\ \cdot & & & & \\ \cdot & & & & \\ 0 & 0 & \cdots & I & 0 \end{bmatrix}$$

acting on $\bigoplus_{k=1}^{n+1} \mathcal{H}_i$, for $i = 1, 2$. We have that $T_n \in \mathcal{I}(S_n(\mathcal{H}_2), S_n(\mathcal{H}_1))$. The next remark is that the minimal isometric dilation of $S_n(\mathcal{H}_1)$ is $S_+(\mathcal{H}_1)$, the shift operator on $\ell^2_+(\mathcal{H}_1)$. Consequently, if $T_\infty \in \mathrm{CID}(T_n)$, as $T_\infty S_+(\mathcal{H}_1) = S_+(\mathcal{H}_2)T_\infty$, it follows that T_∞ is a Toeplitz operator having the symbol $F \in \mathcal{S}(\mathcal{H}_1, \mathcal{H}_2)$. From the other properties of T_∞ as an element in $\mathrm{CID}(T_n)$, we conclude that F is a solution of the Schur problem. In view of Theorem 6.6 it follows that the Schur problem is solvable if and only if $\mathrm{CID}(T_n) \neq \emptyset$, i.e. T_n is a contraction.

Now we consider the following Nevanlinna–Pick problem.

' Given the operators $Z_k \in \mathcal{L}(\mathcal{H}_1, \mathcal{H}_2)$, $k = 1, 2, \cdots, n$ and the distinct complex numbers $\{z_k\}_{k=1}^n \subset \mathbf{D}$, it is required to find conditions for the existence of a function $F \in \mathcal{S}(\mathcal{H}_1, \mathcal{H}_2)$ such that $F(z_k) = Z_k$, $k = 1, \cdots, n$.'

Consider the functions $g_k = 1/(1 - \bar{z}_k z)$, $k = 1, \cdots, n$ and denote by $\mathcal{H}_o$ the subspace of the Hardy space H^2 generated by $\{g_k\}_{k=1}^n$. Then the tensor product Hilbert spaces $\mathcal{H}_o \bigotimes \mathcal{H}_1$ and $\mathcal{H}_o \bigotimes \mathcal{H}_2$ are considered together with the following operators:

$$\begin{aligned} &S_o \in \mathcal{L}(\mathcal{H}_o), \quad S_o^* g_k = \bar{z}_k g_k, \quad k = 1, \cdots, n, \\ &T \in \mathcal{L}(\mathcal{H}_o \bigotimes \mathcal{H}_1), \quad T = S_o \bigotimes I_{\mathcal{H}_1} \\ &T' \in \mathcal{L}(\mathcal{H}_o \bigotimes \mathcal{H}_2), \quad T' = S_o \bigotimes I_{\mathcal{H}_2} \end{aligned}$$

and

$$\begin{aligned} &A : \mathcal{H}_o \bigotimes \mathcal{H}_1 \to \mathcal{H}_o \bigotimes \mathcal{H}_2 \\ &A^*(g_k \bigotimes h_2) = g_k \bigotimes Z_k^* h_2, \quad h_2 \in \mathcal{H}_2, k = 1, \cdots, n. \end{aligned}$$

First, we remark that $A \in \mathcal{I}(T', T)$. Indeed,

$$T^* A^*(g_k \bigotimes h_2) = T^*(g_k \bigotimes Z_k^* h_2) = \bar{z}_k g_k \bigotimes Z_k^* h_2$$

and

$$A^* T'^*(g_k \bigotimes h_2) = A^*(\bar{z}_k g_k \bigotimes h_2) = \bar{z}_k g_k \bigotimes Z_k^* h_2.$$

The next remark is that the minimal isometric dilation of S_o is the shift operator S_+ on H^2 (i.e. $S_+ f(z) = z f(z)$, $f \in H^2$), thus the isometric minimal dilation of T is $U_+ = S_+ \bigotimes I_{\mathcal{H}_1}$ and that of T' is $U'_+ = S_+ \bigotimes I_{\mathcal{H}_2}$. Consequently, take $A_\infty \in \mathrm{CID}(A)$;

then $U'_+ A_\infty = A_\infty U_+$ which implies that A_∞ is a Toeplitz operator with a symbol $F \in \mathcal{S}(\mathcal{H}_1, \mathcal{H}_2)$ and we can rewrite this as $A_\infty = F(S_+)$. As $P_{\mathcal{H}_o \bigotimes \mathcal{H}_2} A_\infty = A P_{\mathcal{H}_o \bigotimes \mathcal{H}_1}$, we get for $h_1 \in \mathcal{H}_1$, $h_2 \in \mathcal{H}_2$, $k = 1, \cdots, n$.

$$\begin{aligned}(A_\infty(g_k \bigotimes h_1), (g_k \bigotimes h_2) &= (g_k \bigotimes F(z_k)h_1, g_k \bigotimes h_2) = (g_k, g_k)(F(z_k)h_1, h_2) \\ &= (A(g_k \bigotimes h_1), g_k \bigotimes h_2) = (g_k \bigotimes h_1, A^*(g_k \bigotimes h_2)) = (g_k \bigotimes h_1, g_k \bigotimes Z_k^* h_2) \\ &= (g_k, g_k)(Z_k h_1, h_2)\end{aligned}$$

i.e. $F(z_k) = Z_k$, $k = 1, \cdots, n$.

We find that the Nevanlinna–Pick problem is solvable if and only if $\mathrm{CID}(A) \neq \emptyset$, that is, by Theorem 6.6, if and only if A is a contraction. But this condition means that for any complex numbers a_k, $k = 1, \cdots, n$ and $h_2 \in \mathcal{H}_2$,

$$0 \leq ((1 - AA^*)(\sum_{k=1}^{n} a_k g_k \bigotimes h_2), \sum_{j=1}^{n} a_j g_j \bigotimes h_2)$$

which, because of the relations $(g_k, g_j) = g_k(z_j) = 1/(1 - z_j \bar{z}_k)$, is equivalent to

$$\sum_{j,k=1}^{n} \frac{a_k \bar{a}_j (I - Z_j Z_k^*)}{1 - z_j \bar{z}_k} \geq 0$$

i.e., the classical criterion for the solvability of the Nevanlinna–Pick problem.

Now, we take into account the Nehari problem with data $\{C_n\}_{n=-\infty}^{-1}$, $C_n \in \mathcal{L}(\mathcal{U}, \mathcal{Y})$. Define $\mathcal{H} = \ell^2_+(\mathcal{U})$, $\mathcal{H}' = \ell^2_+(y)$, $T = S_+(\mathcal{U})$, $T' = S_+^*(\mathcal{Y})$ and the Hankel matrix (viewed as acting between $\mathcal{H}$ and $\mathcal{H}'$)

$$A : \mathcal{H} \to \mathcal{H}'$$

$$A = \begin{bmatrix} C_{-1} & C_{-2} & C_{-3} & \cdots \\ C_{-2} & C_{-3} & \cdots & \\ C_{-3} & \cdots & & \\ \cdot & & & \\ \cdot & & & \\ \cdot & & & \end{bmatrix}$$

Then $A \in \mathcal{I}(T', T)$ and the minimal isometric dilation of T' is $U'^*_+ \in \mathcal{L}(\ell^2(\mathcal{Y}))$, the bilateral shift on $\ell^2(\mathcal{Y})$. Considering $A_\infty \in \mathrm{CID}(A)$ this operator satisfies $U'_+ A_\infty = A_\infty T$; consequently it gives rise to a function f having $\{C_n\}_{n=-\infty}^{-1}$ as its negatively indexed Fourier coefficients. Again by Theorem 6.6, we find that the Nehari problem is solvable if and only if A is a contraction, i.e. the statement of Theorem 6.4.

A last remark concerning the problem formulated at the end of Chapter V. It is a consequence of Theorem 6.4 that for $f \in L^\infty(\mathcal{L}(\mathcal{H}))$, the distance between f and $H^\infty(\mathcal{L}(\mathcal{H}))$ is

$$d(f, H^\infty(\mathcal{L}(\mathcal{H}))) = \| H_f \| \tag{6.66}$$

where H_f is the Hankel operator based on f. Indeed, for any $G \in H^\infty(\mathcal{L}(\mathcal{H}))$,

$$\| H_f \| = \| H_{f+G} \| \leq \| f + G \|_\infty$$

and consequently,

$$\| H_f \| \leq \inf\{\| f + G \| | G \in H^\infty(\mathcal{L}(\mathcal{H})) = d(f, H^\infty(\mathcal{L}(\mathcal{H}))).$$

On the other hand, the above proof of Theorem 6.4 based on the lifting of commutants shows that there exists $g \in L^\infty(\mathcal{L}(\mathcal{H}))$ such that

$$\| g \|_\infty = \| H_f \| \quad \text{and} \quad H_f = H_g.$$

But $H_f = H_g$ means that $G_o = f - g$ belongs to $H^\infty(\mathcal{L}(\mathcal{H}))$ and thus,

$$\inf\{\| f - G \| | G \in H^\infty(\mathcal{L}(\mathcal{H}))\} \leq \| f - G_o \|_\infty = \| g \|_\infty = \| H_f \| .$$

In this way, a solution of the problem is obtained in terms of a Hankel operator.

6.4 Continuous completion problems

We briefly indicate in this section that the results concerning extensions of isometries can also be used to solve so–called continuous completion problems. We consider here only two such problems, those of Hamburger and M. G. Krein.

The Hamburger moment problem has the following statements.

'Given a sequence $\{s_n\}_{n=0}^\infty$ of real numbers, it is required to find conditions for the existence of a positive Borel measure μ on $\mathbf{R}$ such that $s_n = \int t^n d\mu(t)$, $n = 0, 1, \cdots$'

The necessary condition follows easily: let $a_o \cdots, a_n$ be complex numbers, $n = 0, 1, \cdots$; then

$$\sum_{j,k=0}^{n} a_j \bar{a}_k s_{j+k} = \sum_{j,k=0}^{n} a_j \bar{a}_k \int t^{j+k} d\mu(t) = \int \sum_{j,k=0}^{n} a_j \bar{a}_k t^{j+k} d\mu(t)$$
$$= \int | \sum_{j=0}^{n} a_j t^j |^2 \, d\mu(t) \geq 0$$

i.e. the Hankel form $\{s_{j+k}\}_{j,k=0}^{\infty}$ is positive. Conversely, suppose the Hankel form $\{s_{j+k}\}_{j,k=0}^{\infty}$ is positive. Let $\mathcal{P}$ denote the set of polynomials with complex coefficients and define the inner product on $\mathcal{P}$,

$$(\sum_{j=0}^{m} a_j t^j, \sum_{k=0}^{n} b_k t^k) = \sum_{j=0}^{m} \sum_{k=0}^{n} a_j \bar{b}_k s_{j+k}.$$

Renorming $\mathcal{P}$ with this inner product we obtain a Hilbert space $\mathcal{H}$. Define the operator

$$A : \mathcal{P} \to \mathcal{P}$$
$$A(\sum_{j=0}^{m} a_j t^j) = \sum_{j=0}^{m} a_j t^{j+1} \tag{6.67}$$

which satisfies

$$(Ap, q) = (p, Aq) \quad \text{for } p, q \in \mathcal{P}$$

(i.e. A is symmetric), because $\{s_{j+k}\}_{j,k=0}^{\infty}$ is a Hankel form.

The space $\mathcal{H}$ was obtained by completing $\mathcal{P}/\{p \in \mathcal{P} \mid (p,p) = 0\}$ and A extends then to a densely defined symmetric operator on $\mathcal{H}$. Its domain will be $\mathcal{P}/\{p \in \mathcal{P} \mid (p,p) = 0\}$, which is dense in $\mathcal{H}$ and by Schwartz's inequality,

$$(Ap, Ap) = | (A^2 p, p) | \leq (A^2 p, A^2 p)^{\frac{1}{2}} (p, p)^{\frac{1}{2}},$$

so A indeed extends to $\mathcal{P}/\{p \in \mathcal{P} \mid (p,p) = 0\}$ and remains symmetric (this extension is also denoted by A). A being symmetric is closable and its closure is an extension, also denoted by A. We will use the following result.

Proposition 6.15 If A is a closed symmetric operator on $\mathcal{H}$, then there exist self–adjoint extensions of A acting on a Hilbert space $\mathcal{K}$ containing $\mathcal{H}$. (It may happen that $\mathcal{K}$ and $\mathcal{H}$ coincide.)

Proof For the purpose of proving this statement we need to use the Cayley transform. Note that for $g \in D(A)$, the domain of A,

$$\| (A \pm i)g \|^2 = \| Ag \|^2 \pm (Ag, ig) \pm (ig, Ag) + \| g \|^2 = \| Ag \|^2 + \| g \|^2 \geq \| g \|^2 .$$

Consequently, $\mathcal{R}(A \pm i)$ are closed spaces, $A+i$ is injective and we can define the Cayley transform of A by

$$\begin{aligned} V(A) &: \mathcal{R}(A+i) \to \mathcal{R}(A-i) \\ V(A) &= (A-i)(A+i)^{-1} \end{aligned} \tag{6.68}$$

which is a surjective partial isometry.

Let $W \in \mathcal{L}(\mathcal{K})$ be a unitary extension of $V(A)$ given by Theorem 6.1. Moreover, for $g \in D(A)$, $(I - V(A))(A+i)g = (A+i)g - (A-i)g = 2ig$, so that

$$\overline{(I - V(A))\mathcal{R}(A+i)} = \mathcal{H}.$$

Let us show that

$$\overline{(I - W)\mathcal{K}} = \mathcal{K}. \tag{6.69}$$

Take $k \in \mathcal{K}$, $k \perp (I - W)\mathcal{K}$; consequently $k \perp (I - V(A))\mathcal{R}(A+i)$ and $k \perp \mathcal{H}$. But $k \perp (I - W)\mathcal{K}$ implies $k = W^* k$ and, further, $\cdots = W^* k = k = Wk = \cdots$. Then $k \perp \mathcal{H}$ implies $k \perp \bigvee_{n \in \mathbf{Z}} W^n \mathcal{H} = \mathcal{K}$ and $k = 0$. As a consequence of (6.69), we can define

$$\begin{aligned} D(S) &= \mathcal{R}(I - W) \\ S &= i(I + W)(I - W)^{-1}. \end{aligned} \tag{6.70}$$

It is obvious that S is symmetric. To prove that it is closed, take $g_n \in \mathcal{R}(I - W)$, with $g_n \to g$ and $Sg_n \to h$. Consequently there exists a sequence $\{g'_n\} \subset \mathcal{H}$ such that $g_n = (I - W)g'_n$. Having also that $i(I + W)g'_n \to h$ we deduce that $2ig'_n \to h + ig$ and $2iWg'_n \to h - ig = 2iWg'$. Hence, $g'_n \to g'$, $g = (I - W)g' \in D(S)$ and $Sg = S(I - W)g' = i(I + W)g' = ig' + iWg' = h$.

To prove that S is self–adjoint, we take $h \perp \mathcal{R}(S+i)$ and then $(h, (S+i)(I-W)g) = 0$ for $g \in \mathcal{K}$, or, using the definition of S, $(h, g) = 0$ for $g \in \mathcal{K}$, i.e. $h = 0$. In a similar way, $\mathcal{R}(S - i) = \mathcal{K}$. By a standard result, it follows that S is self–adjoint. Obviously, when restricted to $D(A)$, S coincides with A, i.e. it is an extension of A. ∎

Theorem 6.16 The Hamburger moment problem with data $\{s_n\}_{n=1}^{\infty}$ is solvable if and only if the Hankel form $\{s_{i+j}\}$ is positive.

Proof We use Proposition 6.15 for the operator given by (6.67) and let S be a self-adjoint extension of this A. Let E be the spectral measure of S and $\mu(f) = (E(f)1, 1)$, $f \in C(\mathbf{R})$. Then

$$\int t^n d\mu(t) = (S^n 1, 1) = (t^n 1, 1) = s_n. \qquad \blacksquare$$

The last problem we are considering here is that of M. G. Krein:

'Given a continuous function $Q : (-a, a) \to \mathbf{R}$, for $a > 0$, it is required to find conditions for the existence of a positive Borel measure μ on $\mathbf{R}$ such that $Q(x) = \int e^{-ixt} d\mu(t)$.'

The necessary condition can be obtained in the usual manner: for complex numbers $a_o, \cdots, a_n$, $n = 0, 1, \cdots$, and $x_o, \cdots, x_n \in (-a, a)$,

$$\sum_{j,k=0}^{n} a_j \bar{a}_k Q(x_j - x_k) = \sum_{j,k=0}^{n} a_j \bar{a}_k \int e^{-i(x_k - x_j)t} d\mu(t)$$
$$= \int |\sum_{j=0}^{n} a_j e^{ix_j t}|^2 \, d\mu(t) \geq 0$$

i.e. the function Q is positive definite on $(-a.a)$.

Our proof that this condition is also sufficient is modelled after a proof of Bochner's theorem and requires some elements of generalised semigroups of contractions.

A family $\{T_t\}_{t \geq 0}$ of contractions on a Hilbert space $\mathcal{H}$ is a generalised semigroup of contractions if

(a) The domains $D(T_t) = \mathcal{E}_t$ are closed subspaces of $\mathcal{H}$ and

$$\mathcal{E}_t \subset \mathcal{E}_{t'} \quad \text{for } 0 \leq t' \leq t,$$
$$\overline{\bigcup_{t>0} \mathcal{E}_t} = \mathcal{H}, \quad T_{t'} \mathcal{E}_{t'+t} \subset \mathcal{E}_t.$$

(b) $T_o = I_{\mathcal{H}}$, $T_{t+t'} = T_t T_{t'}$, $t, t' \geq 0$.

(c) If $t_o > 0$, $h \in \mathcal{E}_{t_o}$ and $0 \leq t, t' \leq t_o$, then $\lim_{t' \to t} T_{t'} h = T_t h$.

Several results in the theory of semigroups of contractions remain true in this more general context. Thus, define the infinitesimal generator by

$$D(A) = \{h \in \bigcup_{t>0} \mathcal{E}_t, \lim_{t\downarrow 0} \frac{T_t h - h}{t} \text{ exists}\}$$

$$Ah = \lim_{t\downarrow 0} \frac{T_t h - h}{t} \quad \text{for } h \in D(A). \tag{6.71}$$

Proposition 6.17

(i) If $h \in D(A) \cap \mathcal{E}_t$, then $T_s h \in D(A)$, $s > 0$, $Ah \in \mathcal{E}_t$ and $(d/dt)T_t h = AT_t h = T_t Ah$.

(2) If $h \in \mathcal{E}_{t_o}$, $0 < t < t_o$, then $\int_0^t T_s h ds \in D(A)$ and $A \int_0^t T_s h ds = T_t h - h$.

(3) $D(A)$ is dense in $\mathcal{H}$.

(4) Defining $B = -iA$ and supposing T_t are isometries for $t > 0$, then B is symmetric.

Proof

(1) Take $h \in \mathcal{E}_t \cap D(A)$, $s > 0$. Then $h \in \mathcal{E}_{s+t}$, $T_s \mathcal{E}_{s+t} \subset \mathcal{E}_t$, $T_t \mathcal{E}_{t+s} \subset \mathcal{E}_s$ and

$$\frac{T_{s+t}h - T_t h}{s} = T_t \frac{T_s h - h}{s} = \frac{T_s - I}{s} T_t h.$$

But $h \in D(A)$; consequently $T_t h \in D(A)$, $Ah \in \mathcal{E}_t$ and by letting $s \to 0$. $T_t Ah = AT_t h = (d/dt)T_t h$.

(2) $h \in \mathcal{E}_{t_o}$, so $h \in \mathcal{E}_t$ for $0 < t < t_o$. Define $h_t = \int_0^t T_s h ds$ and take $r > 0$ such that $t + r < t_o$; then

$$\begin{aligned} \frac{T_r h_t - h_t}{r} &= \frac{1}{r}\left(\int_t^{t+r} T_s h ds - \int_0^t T_s h ds\right) \\ &= \frac{1}{r}\int_t^{t+r} T_s h ds - \frac{1}{r}\int_0^t T_s h ds \underset{r\downarrow 0}{\to} T_t h - h. \end{aligned}$$

Consequently $\int_0^t T_s h ds \in D(A)$, $A \int_0^t T_s h ds = T_t h - h$.

(3) Take $h \in \mathcal{H}$; then there exists $h' \in \bigcup_{t>0} \mathcal{E}_t$ with $\| h - h' \| < \epsilon$. $h' \in \mathcal{E}_{t_o}$ for a certain $t_o > 0$ and according to (2), $h'_t \in D(A)$ for $0 < t < t_o$ and since $(1/t)h'_t \underset{t \downarrow 0}{\rightarrow} h'$, it follows that $\overline{D(A)} = \mathcal{H}$.

(4) Take $h, g \in D(A)$, $t_o > 0$, $h, g \in \mathcal{E}_{t_o}$; then for $0 < t < t_o$,

$$(\frac{T_t - I}{t} h, g) = (T_t \frac{T_t - I}{t} h, T_t g) = (T_t h, \frac{g - T_t g}{t})$$

and letting $t \downarrow 0$, it follows that $(Bh, g) = (h, Bg)$. ∎

Theorem 6.18 The Krein problem is solvable if and only if Q is positive definite.

Proof Consider $\mathcal{F}$ to be the set of functions defined on $(-a, a)$ with values in $\mathbf{C}$ and with finite support. Define the inner product

$$(f, g) = \sum_{x,y} Q(x - y) f(x) \overline{g(y)} \quad f, g \in \mathcal{F}, x, y \in (-a, a).$$

Renorming $\mathcal{F}$ with this inner product, we get a Hilbert space and define further

$$\mathcal{E}_t = \{ f \in \mathcal{F} \mid t + \operatorname{supp} f \subset (-a, a) \}$$

and for $f \in \mathcal{E}_t$,

$$V_t f(s) = \begin{cases} f(s - t) & \text{when } -a + t \leq s < a \\ 0 & \text{when } -a < s \leq -a + t \end{cases}$$

which are isometries. Furthermore, we verify that $\{V_t\}_{t \geq 0}$ is a generalised semigroup of isometries. Let A be its infinitesimal generator and $B = -iA$, which is symmetric by Proposition 6.17. It may be that B is not closed, but being symmetric it is closable and we must take into account its closure, denoted also by B.

Considering, by Proposition 6.15, a self–adjoint extension C of B, it will produce in an obvious way a group $\{U_t\}_{t \in \mathbf{R}}$ of unitary operators on a Hilbert space $\mathcal{K}$ containing $\mathcal{H}$ such that $V_t = U_t \mid \mathcal{E}_t$ for $t \geq 0$. Let E be the spectral measure of C and define

$$f_o(x) = \begin{cases} 1, & x = 0 \\ 0, & x \neq 0 \end{cases}$$

which belongs to $\mathcal{F}$. Then $\mu(f) = (E(f) f_o, f_o)$, $f \in C(\mathbf{R})$, will be a solution of the problem. Indeed,

$$Q(x) = (V_x f_o, f_o) = (U_x f_o, f_o) = (E(e^{-ix}) f_o, f_o) = \mu(e^{-ix})$$

$$= \int e^{-itx} d\mu(t). \qquad \blacksquare$$

Notes

The idea of using extensions of isometries for solving moment problems was first developed by Naimark [66]. The first generalised resolvent formula appeared in [57] and [58]. Theorem 6.1 and 6.3 were proved in [26]. For applications to other moment problems see, for instance, [6] and [29].

The Nehari problem was solved in [67] and then developed in [1] and [2]. The approach os using the lifting of commutants was indicated in [1] and [68]. The theory of contractive intertwining dilations starts with the lifting theorem of Sarason [76] and Sz. Nagy and Foiaş [84] and was developed in [4], [20], [21] and [9].

The first proof of Theorem 6.6 presented here is in [6]. The rest of the material in Section 6.3, especially Theorems 6.12 and 6.13, and Remark 6.14 are in [9]. We used here a certain variant presented in [25]. The fact that the Schur problem and the Nevanlinna–Pick problem are lifting of commutant problems was shown in [76]. For the continuous problems in Section 6.4 we used [59], [60] and [77]. Other papers concerned with completion problems are [39], [12] and [69].

APPENDIX

Some positive and contractive structures

In this Appendix some simple block matrices are described in detail. These are basic objects in dilation theory and are repeatedly used in this book. We begin with a useful result on the positivity of a 2×2 block matrix. Fix a positive operator $A \in \mathcal{L}(\mathcal{H})$ and consider an extension of A, $H = \begin{bmatrix} A & B \\ B^* & C \end{bmatrix}$ on the Hilbert space $\mathcal{H} \bigoplus \mathcal{H}'$.

Theorem A.1 Let $A \in \mathcal{L}(\mathcal{H})$, $A \geq 0$ and H as above. Supposing C is a fixed positive operator, H is positive if and only if there exists a contraction $G \in \mathcal{L}(\overline{\mathcal{R}(C)}, \overline{\mathcal{R}(A)})$ such that $B = A^{\frac{1}{2}} G C^{\frac{1}{2}}$.

Proof The first step is to suppose A and C are invertible. Indeed, the following Frobenius–Schur identity holds

$$\begin{bmatrix} X & Y \\ Z & U \end{bmatrix} = \begin{bmatrix} I & 0 \\ ZX^{-1} & I \end{bmatrix} \begin{bmatrix} X & 0 \\ 0 & U - ZX^{-1}Y \end{bmatrix} \begin{bmatrix} I & X^{-1}Y \\ 0 & I \end{bmatrix}$$

for arbitrary operators X, Y, Z, U and X invertible. Using this formula in our case, H is positive if and only if $C - B^*A^{-1}B$ is positive. Therefore, $I - C^{-\frac{1}{2}} B^* A^{-\frac{1}{2}} A^{-\frac{1}{2}} B C^{-\frac{1}{2}} \geq 0$ and with $G = A^{-\frac{1}{2}} B C^{-\frac{1}{2}}$, we obtain $\| G \| \leq 1$ and $B = A^{\frac{1}{2}} G C^{\frac{1}{2}}$.

For the general case, for every $n \geq 1$, $A_n = A + (1/n)I$, $C_n = C + (1/n)I$ are positive and invertible operators so we find contractions G_n with $B = A_n^{\frac{1}{2}} G_n C_n^{\frac{1}{2}}$ for $n \geq 1$. From the weak compactness of the unit ball in $\mathcal{L}(\mathcal{H})$, a limit point of the sequence $\{G_n\}_{n \geq 1}$ in the weak topology will be a contraction G satisfying $B = A^{\frac{1}{2}} G C^{\frac{1}{2}}$.

Finally, if G' is another contraction in $\mathcal{L}(\overline{\mathcal{R}(C)}, \overline{\mathcal{R}(A)})$ with $B = A^{\frac{1}{2}} G' C^{\frac{1}{2}}$, then $A^{\frac{1}{2}} G C^{\frac{1}{2}} h = A^{\frac{1}{2}} G' C^{\frac{1}{2}} h$ for $h \in \mathcal{H}'$ and as A is injective on the closure of its range, $G C^{\frac{1}{2}} h = G' C^{\frac{1}{2}} h$ for $h \in \mathcal{H}'$, that is $G' = G$ on $\overline{\mathcal{R}(C)}$. ■

A first consequence of Theorem A.1 is an equally useful factorisation result.

Proposition A.2 For two fixed operators A and B in $\mathcal{L}(\mathcal{H})$ there exists a contraction $G \in \mathcal{L}(\mathcal{H})$ such that $A = GB$ if and only if

$$A^*A \leq B^*B.$$

When A and B have dense ranges, then G above is unitary if and only if $A^*A = B^*B$.

Proof Suppose that $A^*A \leq B^*B$. Then

$$\begin{bmatrix} I & A \\ A^* & B^*B \end{bmatrix} \geq 0$$

and by Theorem A.1, $A = G'(B^*B)^{\frac{1}{2}}$ with G' a contraction. Take the polar decomposition $B = V(B^*B)^{\frac{1}{2}}$ and $A = G'V^*B$ with $G = G'V$ a contraction. ∎

Another application of the factorisation (A.1) gives a formula for the inverse of a 2×2 block matrix.

Proposition A.3 Consider $T = \begin{bmatrix} X & Y \\ Z & U \end{bmatrix}$ with X an invertible operator. Then T is invertible if and only if the Schur complement $S = U - ZX^{-1}Y$ is invertible and then

$$T^{-1} = \begin{bmatrix} X^{-1} + X^{-1}YS^{-1}ZX^{-1} & -X^{-1}YS^{-1} \\ -S^{-1}ZX^{-1} & S^{-1} \end{bmatrix}.$$

We describe next the structure of all contractions of the form

$$T^{(n)} = (T_1, T_2, \cdots, T_n) : \mathcal{H}(= \bigoplus_{k=1}^{n} \mathcal{H}_k) \to \mathcal{H}'.$$

Theorem A.4 $T^{(n)}$ is a contraction if and only if $G_1 = T_1$ is a contraction, and $T_k = D_{G_1^*} \cdots D_{G_{k-1}^*} G_k$, $k \geq 2$, where $G_k : \mathcal{H}_k \to \mathcal{D}_{G_{k-1}^*}$ are contractions.

The correspondence between $T^{(n)}$ and $\{G_k\}_{k=1}^n$ is one to one and the identification of the defect spaces of $T^{(n)}$ can be explicitly given by the following unitary operators:

$$\alpha_n : \mathcal{D}_{T^{(n)}} \to \mathcal{D}_{G_1} \bigoplus \mathcal{D}_{G_2} \bigoplus \cdots \bigoplus \mathcal{D}_{G_n}$$

$$\alpha_n D_{T^{(n)}} = \begin{bmatrix} D_{G_1} & -G_1^*G_2 & -G^*D_{G_2^*}G_3 & \cdots & -G_1^*D_{G_2^*} \cdots D_{G_{n-1}^*}G_n \\ 0 & D_{G_2} & -G_2^*G_3 & \cdots & -G_2^*D_{G_3^*} \cdots D_{G_{n-1}^*}G_n \\ \cdot & & & & \cdot \\ \cdot & & & & \cdot \\ \cdot & & & & \cdot \\ 0 & & & \cdots & D_{G_n} \end{bmatrix}$$

and

$$\beta_n : \mathcal{D}_{T^{(n)*}} \to \mathcal{D}_{G_n^*}$$
$$\beta_n D_{T^{(n)*}} = D_{G_n^*} \cdots D_{G_1^*}.$$

Proof For the case $n = 2$, we use Theorem A.1 twice, first to obtain

$$\begin{bmatrix} I & G_1 & T_2 \\ G_1^* & I & 0 \\ T_2^* & 0 & I \end{bmatrix} \geq 0$$

and then to write

$$\begin{bmatrix} T_2 \\ 0 \end{bmatrix} = \begin{bmatrix} I & G_1 \\ G_1^* & I \end{bmatrix}^{\frac{1}{2}} \cdot G_2'$$

where G_2' is a contraction. On the basis of the Frobenius–Schur factorisation (A.1), it follows that

$$\begin{bmatrix} T_2 \\ 0 \end{bmatrix} = \begin{bmatrix} D_{G_1^*} & G_1 \\ 0 & I \end{bmatrix} \begin{bmatrix} (G_2'')_1 \\ (G_2'')_2 \end{bmatrix} = \begin{bmatrix} D_{G_1^*}(G_2'')_1 + G_1(G_2'')_2 \\ (G_2'')_2 \end{bmatrix}$$

where $G_2'' = \begin{bmatrix} (G_2'')_1 \\ (G_2'')_2 \end{bmatrix}$ is another contraction. Consequently, $(G_2'')_2 = 0$ and we can take $G_2 = (G_2'')_1$. Moreover, by (A.1), the following factorisation holds

$$\begin{bmatrix} I & G_1 & T_2 \\ G_1^* & I & 0 \\ T_2^* & 0 & I \end{bmatrix} = \begin{bmatrix} I & 0 & 0 \\ G_1^* & D_{G_1} & 0 \\ G_2^* D_{G_1^*} & -G_2^* G_1 & D_{G_2} \end{bmatrix} \begin{bmatrix} I & G_1 & D_{G_1^*} G_2 \\ 0 & D_{G_1} & -G_1^* G_2 \\ 0 & 0 & D_{G_2} \end{bmatrix}$$

and the unitary operators

$$\alpha_2 : \mathcal{D}_{T^{(2)}} \to \mathcal{D}_{G_1} \bigoplus \mathcal{D}_{G_2}$$
$$\alpha_2 D_{T^{(2)}} = \begin{bmatrix} D_{G_1} & -G_1^* G_2 \\ 0 & D_{G_2} \end{bmatrix}$$

and

$$\beta_2 : \mathcal{D}_{T^{(2)*}} \to \mathcal{D}_{G_2^*}$$
$$\beta_2 D_{T^{(2)*}} = D_{G_2^*} D_{G_1^*}$$

are naturally imposed. For the general case we proceed by induction. Suppose that for every row–contraction with $n - 1$ entries

$$T^{(n-1)} = (G_1, D_{G_1^*} G_2, \cdots, D_{G_1^*} \cdots D_{G_{n-2}^*} G_{n-1})$$

for certain contractions $G_1 \in \mathcal{L}(\mathcal{H}_1, \mathcal{H}')$ and $G_k \in \mathcal{L}(\mathcal{H}_k, \mathcal{D}_{G_{k-1}})$, $2 \le k \le n-1$, together with the corresponding identifications of the defect spaces. Then $T^{(n)} = ((T_1, \cdots, T_{n-1}), T_n)$ is a contraction if and only if $T^{(n-1)} = (T_1, \cdots, T_{n-1})$ is a contraction and $T_n = D_{T^{(n-1)*}} G'_n$, where $G'_n : \mathcal{H}_n \to \mathcal{D}_{T^{(n-1)*}}$ is a uniquely determined contraction. But, by the induction hypothesis, $T^{(n-1)} = T^{(n-1)}(G_1, \cdots, G_{n-1})$ for a certain family $\{G_k\}_{k=1}^{n-1}$ and, defining $G_n : \mathcal{H}_n \to \mathcal{D}_{G^*_{n-1}}$ by $G_n = \beta_{T^{(n-1)}(G_1, \cdots, G_{n-1})} G'_n$, we get $T_n = D_{G_1^*} \cdots D_{G^*_{n-1}} G_n$. Moreover, $D^2_{T^{(n)*}} = D_{G_1^*} \cdots D^2_{G_n^*} \cdots D_{G_1^*}$ and

$$\beta_n : \mathcal{D}_{T^{(n)*}} \to \mathcal{D}_{G_n^*}$$
$$\beta_n D_{T^{(n)*}} = D_{G_n^*} \cdots D_{G_1^*}.$$

is a unitary operator. For the identification $\mathcal{D}_{T^{(n)}}$, we write $T^{(n)} = (G_1, D_{G_1^*} T^{(n-1)}(G_2, \cdots G_n))$ and α_n is given by the following diagram:

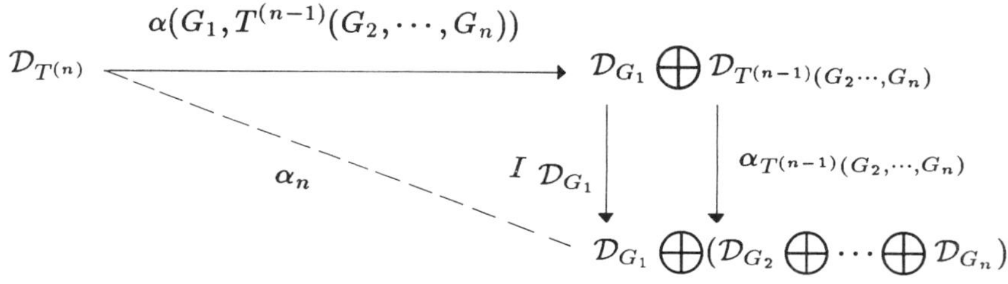

∎

For this kind of contraction we use the special notation $L_n = L_n(\{G_k\}_{k=1}^n) = (G_1, D_{G_1^*} G_2, \cdots, D_{G_1^*} \cdots D_{G^*_{n-1}} G_n)$, when the parameters are specified from the beginning and $L(\{G_k\}_{k=1}^n)$ when this specification is necessary. For the unitary operators identifying the defect spaces of L_n we keep the notation α_n and β_n, and when necessary we write $\alpha(\{G_k\}_{k=1}^n)$ and $\beta(\{G_k\}_{k=1}^n)$. One more notational convention is maintained throughout. The operators in the right–hand side of the definition of $\alpha(\{G_k\}_{k=1}^n)$ are denoted by $D(\{G_k\}_{k=1}^n)$ (or simply D_n when this is possible) and the ones in the right–hand side of the definition of $\beta(\{G_k\}_{k=1}^n)$ are denoted by $H(\{G_k\}_{k=1}^n)$ (or H_n). Another useful notation is produced by the following duality: for a formula form $(\{G_k\}_{k=1}^n)$ depending on the parameters $\{G_k\}_{k=1}^n$, we define

$$\widetilde{\text{form}}(\{G_k\}_{k=1}^n) = \text{form}(\{G_k^*\}_{k=1}^n)^*.$$

It is obvious that ~ is a duality. From Therem A.4 we get the structure of the column–contractions $\tilde{L}(\{G_k\}_{k=1}^n) = (G_1, G_2 D_{G_1}, \cdots, G_n D_{G_{n-1}} \cdots D_{G_1})^t : \mathcal{H} \to \mathcal{H}'(= \bigoplus_{k=1}^{n} \mathcal{H}'_k)$ with G_1 a contraction in $\mathcal{L}(\mathcal{H}, \mathcal{H}'_1)$ and $G_k : \mathcal{D}_{G_{k-1}} \to \mathcal{H}'_k$ also contractions for $k \geq 2$ (the sign 't' is used for matrix transpose).

The identifications of the defect spaces are obtained from Theorem A.4 by means of the unitary operators $\tilde{\alpha}(\{G_k\}_{k=1}^n) : \mathcal{D}_{\tilde{L}(\{G_k\}_{k=1}^n)^*} \to \mathcal{D}_{G_1^*} \bigoplus \mathcal{D}_{G_2^*} \bigoplus \cdots \bigoplus \mathcal{D}_{G_n^*}$ and $\tilde{\beta}(\{G_k\}_{k=1}^n) : \mathcal{D}_{\tilde{L}(\{G_k\}_{k=1}^n)} \to \mathcal{D}_{G_n}$.

We now analyse row–contractions of infinite length. An application of Theorem A.4 shows that any such contraction has the structure:

$$L = L(\{G_n\}_{n\geq 1}) = (G_1, D_{G_1^*} G_2, \cdots) : \mathcal{H}(= \bigoplus_{n=1}^{\infty} \mathcal{H}_n) \to \mathcal{H}'$$

with uniquely determined contractions $\{G_n\}_{n\geq 1}$, $G_1 \in \mathcal{L}(\mathcal{H}_1, \mathcal{H}')$, $G_n \in \mathcal{L}(\mathcal{H}_n, \mathcal{D}_{G_{n-1}})$, $n \geq 2$. We are faced now with the problem of identifying the defect spaces of L. For the projection of $\mathcal{H}$ onto $\bigoplus_{k=1}^{n} \mathcal{H}_k$ we use the notation P_n, and $\mathrm{s} - \lim_{n\to\infty}$ denotes the strong operatorial limit.

Lemma A.5 The following strong operatorial limits exist and define bounded operators:

$$D_\infty(L) : \mathcal{H} \to \bigoplus_{n=1}^{\infty} \mathcal{D}_{G_n}$$
$$D_\infty(L) = \mathrm{s} - \lim_{n\to\infty} D(\{G_k\}_{k=1}^n) P_n$$

and

$$H_\infty(L) : \mathcal{H}' \to \mathcal{H}'$$
$$H_\infty(L) = \mathrm{s} - \lim_{n\to\infty} H(\{G_k\}_{k=1}^n)^* H(\{G_k\}_{k=1}^n))^{\frac{1}{2}}.$$

Proof For $h \in \bigcup_{n=1}^{\infty} \bigoplus_{k=1}^{n} \mathcal{H}_k$ there exists $p \in \mathbf{N}$ such that $h \in \bigoplus_{k=1}^{p} \mathcal{H}_k$ and for $n \geq p$, $D(\{G_k\}_{k=1}^n)h = D(\{G_k\}_{k=1}^p)h$; consequently, the sequence $\{D(\{G_k\}_{k=1}^n)P_n\}_{n\geq 1}$ is a Cauchy sequence. But $\| D(\{G_k\}_{k=1}^n) \| \leq 1$ and $\overline{\bigcup_{n=1}^{\infty} \bigoplus_{k=1}^{n} \mathcal{H}_k} = \mathcal{H}$, so there exists $D_\infty(L) = \mathrm{s} - \lim_{n\to\infty} D(\{G_k\}_{k=1}^n)P_n$ and it is a contraction.

For the second part, we remark that

$$\begin{aligned} & H(\{G_k\}_{k=1}^n)^* H(\{G_k\}_{k=1}^n) \\ & = H(\{G_k\}_{k=1}^{n-1})^* H(\{G_k\}_{k=1}^{n-1}) - (D_{G_1^*} \cdots D_{G_{n-1}^*} G_n)(D_{G_1^*} \cdots D_{G_{n-1}^*} G_n)^* \end{aligned}$$

which shows that $\{H(\{G_k\}_{k=1}^n)^* H(\{G_k\}_{k=1}^n)\}_{n\geq 1}$ is a monotone bounded sequence of positive operators. Consequently, there exists

$$H_\infty(L) = (\mathrm{s} - \lim_{n\to\infty} H(\{G_k\}_{k=1}^n)^* H(\{G_k\}_{k=1}^n)$$

and it defines a contraction. ∎

Proposition A.6 There exists the unitary operators

$$\begin{aligned} \alpha(L) : \mathcal{D}_L &\to \bigoplus_{n=1}^{\infty} \mathcal{D}_{G_n} = \mathcal{D}(L) \\ \alpha(L) D_L &= D_\infty(L) \end{aligned}$$

and

$$\begin{aligned} \beta(L) : \mathcal{D}_{L^*} &\to \overline{\mathcal{R}(H_\infty(L))} = \mathcal{D}_*(L) \\ \beta(L) D_{L^*} &= H_\infty(L). \end{aligned}$$

Proof It is easy to see that

$$\mathrm{s} - \lim_{n\to\infty} L_n P_n = L$$

and

$$\mathrm{s} - \lim_{n\to\infty} L_n^* = L^*$$

(L_n^* is viewed as acting between $\mathcal{H}'$ and $\mathcal{H}$). It follows that $\| D_{L_n P_n} h \| \to \| D_L h \|$ for $h \in \mathcal{H}$ and, in view of Lemma A.5, $\| D_n P_n h \| \to \| D_\infty h \|$ for $h \in \mathcal{H}$ (with an obvious meaning for D_n). But by Theorem A.4, $\| D_{L_n P_n} h \| = \| D_n P_n h \|$ for $h \in \mathcal{H}$ and we deduce that $\alpha(L)$ is a unitary operator.

To prove the unitarity of $\beta(L)$, we remark that

$$(LP_n)(LP_n)^* + H_n^* H_n = I \quad \text{for } n \geq 1$$

(where, of course, $H_n = H(\{G_k\}_{k=1}^n)$). As both sequences in the left–hand side have strong operatorial limits when $n \to \infty$, we obtain

$$LL^* + H_\infty^2 = I.$$

From this equality, $\beta(L)$ is a unitary operator. ∎

The operators L_2 and $\tilde{L}_2$ can be used for obtaining the structure of an $m \times n$ contractive block matrix. However, it is also useful to isolate the 2×2 case.

Theorem A.7 $T = \begin{bmatrix} A & B \\ C & D \end{bmatrix} : \mathcal{H} = \mathcal{H}_1 \bigoplus \mathcal{H}_2 \to \mathcal{H}' = \mathcal{H}_1' \bigoplus \mathcal{H}_2'$ is a contraction if and only if A is a contraction, $B = D_{A^*} G_1$, $C = G_2 D_A$ and $D = -G_2 A^* G_1 + D_{G_2^*} G D_{G_1}$, where the contractions $G_1 : \mathcal{H}_2 \to \mathcal{D}_{A^*}$, $G_2 : \mathcal{D}_A \to \mathcal{H}_1'$ and $G : \mathcal{D}_{G_1} \to \mathcal{D}_{G_2^*}$ are uniquely determined by T. The defect spaces of T are identified by the following unitary operators:

$$\omega_T : \mathcal{D}_T \to \mathcal{D}_{G_2} \bigoplus \mathcal{D}_G$$
$$\omega_T D_T = \begin{bmatrix} D_{G_2} D_A & -(D_{G_2} A^* G_1 + G_2^* G D_{G_1}) \\ 0 & D_G D_{G_1} \end{bmatrix}$$

and

$$\tilde{\omega}_T : \mathcal{D}_{T^*} \to \mathcal{D}_{G_1^*} \bigoplus \mathcal{D}_{G^*}$$
$$\tilde{\omega}_T = \omega_{\tilde{T}}.$$

Proof As $(A, C)^t$ and (A, B) are contractions, C and B have the required form by Theorem A.4. Moreover, we can write

$$(B, D)^t = D_{(A^*, C^*)} G'$$

with G' a contraction. Using the identification of $D_{(A^*, C^*)}$ in Theorem A.4, we get

$$\begin{bmatrix} B \\ D \end{bmatrix} = \begin{bmatrix} D_{A^*} & 0 \\ -G_2 A^* & D_{G_2^*} \end{bmatrix} \begin{bmatrix} G_1'' \\ G_2'' \end{bmatrix} = \begin{bmatrix} D_{A^*} G_1'' \\ -G_2 A^* G_1'' + D_{G_2^*} G_2'' \end{bmatrix}.$$

Consequently $G_1'' = G_1$ and, as $G_2'' = GD_{G_1}$, we obtain

$$D = -G_2A^*G_1 + D_{G_2^*}GD_{G_1}.$$

For the identification of the defect space $\mathcal{D}_T$, we use the diagram

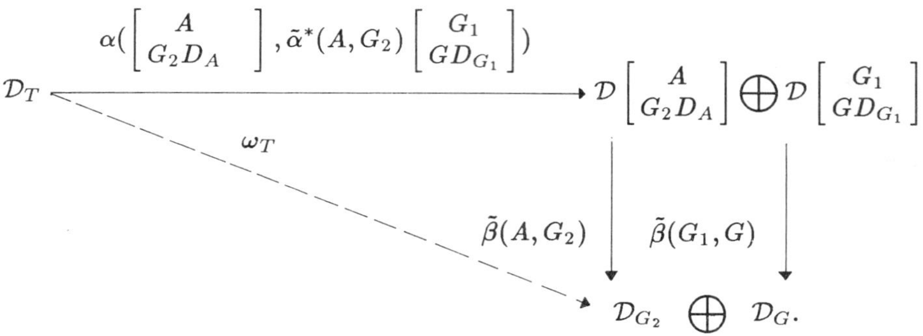

For a given contraction $T \in \mathcal{L}(\mathcal{H},\mathcal{H}')$, the unitary operator (called the elementary rotation of T),

$$J(T) : \mathcal{H} \bigoplus \mathcal{D}_{T^*} \to \mathcal{H}' \bigoplus \mathcal{D}_T$$

$$J(T) = \begin{bmatrix} T & D_{T^*} \\ D_T & -T^* \end{bmatrix}$$

plays a special role, being a fundamental cell for dilation theory.

We conclude this Appendix with some computations regarding the structure of the elementary rotations of the row–contractions $L_n = L(\{G_k\}_{k=1}^n)$ and $L = L(\{G_n\}_{n=1}^\infty)$. Having in mind the identifications of the defect spaces of L_n it is natural to take into account the operators:

$$V_n : \mathcal{H} \bigoplus \mathcal{D}_{G_1} \bigoplus \cdots \bigoplus \mathcal{D}_{G_{n-1}} \bigoplus \mathcal{D}_{G_n^*} \to \mathcal{H}' \bigoplus \mathcal{D}_{G_1} \bigoplus \cdots \bigoplus \mathcal{D}_{G_n}$$

$$V_n = \begin{bmatrix} I & 0 \\ 0 & \alpha_n \end{bmatrix} J(L_n) \begin{bmatrix} I & 0 \\ 0 & \beta_n^* \end{bmatrix}.$$

The only unknown element in the matrix of V_n is $K_n = \alpha_n L_n^* \beta_n^*$.

Lemma A.8 $K_n = (G_1^* D_{G_2^*} \cdots D_{G_n^*}, \cdots, G_{n-1}^* D_{G_n^*}, G_n^*)^t$.

Proof The elements $D_{G_n^*}\cdots D_{G_1^*}h$, $h\in\mathcal{H}$ form a dense part of $\mathcal{D}_{G_n^*}$ and

$$\alpha_n L_n^*\beta_n^* D_{G_n^*}\cdots D_{G_1^*} = \alpha_n L_n^* D_{L_n^*} = \alpha_n D_{L_n} L_n^* = D_n L_n^*.$$

Then, the operators $D_n L_n^*$ can be written as

$$\begin{aligned}(D_nL_n^*)_k &= D_{G_k}G_k^*D_{G_{k-1}^*}\cdots D_{G_1^*} - G_k^*G_{k+1}G_{k+1}^*D_{G_k^*}\cdots D_{G_1^*}\\ &- \sum_{p=1}^{n-k-1} G_k^*D_{G_{k+1}^*}\cdots D_{G_{k+p}^*}G_{k+p+1}G_{k+p+1}^*D_{G_{k+p}^*}\cdots D_{G_1^*}\\ &= G_k^*D_{G_{k+1}^*}\cdots D_{G_n^*}^2 D_{G_{n-1}^*}\cdots D_{G_1^*} \quad \text{for } 1\le k<n\end{aligned}$$

and

$$(D_nL_n^*)_n = D_{G_n}G_n^*D_{G_{n-1}^*}\cdots D_{G_1^*} = G_n^*D_{G_n^*}\cdots D_{G_1^*}.$$ ■

To simplify the notation we define $G_0 = 0 : \mathcal{H}\to\mathcal{H}'$. Then we define the unitary operators:

$$\begin{aligned}J_n(G_k) :& \bigoplus_{j=1}^{k-1}\mathcal{D}_{G_{j-1}}\bigoplus(\mathcal{D}_{G_{k-1}}\bigoplus\mathcal{D}_{G_k^*})\bigoplus\bigoplus_{j=k+1}^{n}\mathcal{D}_{G_j}\\ \to& \bigoplus_{j=1}^{k-1}\mathcal{D}_{G_{j-1}}\bigoplus(\mathcal{D}_{G_{k-1}^*}\bigoplus\mathcal{D}_{G_k})\bigoplus\bigoplus_{j=k+1}^{n}\mathcal{D}_{G_j} \quad \text{for } 1\le k<n\end{aligned}$$

and

$$J_n(G_k) = I\bigoplus J(G_k)\bigoplus I$$

which enter into the multiplicative structure of V_n.

Proposition A.9 $V_n = J_n(G_1)\cdots J_n(G_n)$.

Proof For $n=1$ the equality is obvious, so we suppose it is true up to an $n\in\mathbf{N}$. Then,

$$\begin{aligned}J_{n+1}(G_1)\cdots J_{n+1}(G_{n+1}) &= \begin{bmatrix} V_n & 0\\ 0 & I\end{bmatrix}\begin{bmatrix} I_n & 0\\ 0 & J(G_{n+1})\end{bmatrix}\\ &= \begin{bmatrix} L_n & D_{G_1^*}\cdots D_{G_n^*} & 0\\ D_n & -K_n & 0\\ 0 & 0 & I\end{bmatrix}\begin{bmatrix} I_n & 0 & 0\\ 0 & G_{n+1} & D_{G_{n+1}^*}\\ 0 & D_{G_{n+1}} & -G_{n+1}^*\end{bmatrix}\\ &= \begin{bmatrix} L_n & D_{G_1^*}\cdots D_{G_n^*}G_{n+1} & D_{G_1^*}\cdots D_{G_{n+1}^*}\\ D_n & -K_nG_{n+1} & -K_nD_{G_{n+1}^*}\\ 0 & D_{G_{n+1}} & -G_{n+1}^*\end{bmatrix}.\end{aligned}$$

But

$$\begin{bmatrix} D_n & -K_n G_{n+1} \\ 0 & D_{G_{n+1}} \end{bmatrix} = D_{n+1}$$

and

$$\begin{bmatrix} K_n D_{G^*_{n+1}} \\ G^*_{n+1} \end{bmatrix} = K_{n+1}$$

and these complete the proof. ■

Finally, we consider a row–contraction $L = L(\{G_n\}_{n=1}^{\infty})$: $\mathcal{H} = \bigoplus_{n=1}^{\infty} \mathcal{H}_n \to \mathcal{H}'$ and the unitary operator

$$V_\infty : \mathcal{H} \bigoplus \mathcal{D}_*(L) \to \mathcal{H}' \bigoplus \mathcal{D}(L)$$
$$V_\infty = \begin{bmatrix} I & 0 \\ 0 & \alpha(L) \end{bmatrix} J(L) \begin{bmatrix} I & 0 \\ 0 & \beta^*(L) \end{bmatrix}.$$

Again, it remains to compute only $K_\infty = \alpha(L)L^*\beta^*(L) : \mathcal{D}_*(L) \to \mathcal{D}(L)$ and let us define

$$Z_k : \mathcal{D}_*(L) \to \mathcal{D}_{G_k}$$
$$Z_k = P_{\mathcal{D}_{G_k}}^{\mathcal{H}' \oplus \mathcal{D}(L)} K_\infty.$$

In order to determine the explicit structure of Z_k we take $h \in \mathcal{H}$ and we have

$$\begin{aligned} Z_k H_\infty(L)h &= P_{\mathcal{D}_{G_k}}^{\mathcal{H}' \oplus \mathcal{D}(L)} \alpha(L)L^*\beta^*(L)H_\infty(L)h \\ &= P_{\mathcal{D}_{G_k}}^{\mathcal{H}' \oplus \mathcal{D}(L)} \alpha(L)L^* D_{L^*} h = P_{\mathcal{D}_{G_k}}^{\mathcal{H}' \oplus \mathcal{D}(L)} D_\infty L^* h \\ &= G_k^*(I - L(\{G_m\}_{m=k+1}^{\infty})L(\{G_m\}_{m=k+1}^{\infty})^*)D_{G_k^*} \cdots D_{G_1^*} h \\ &= G_k^* H_\infty^2(L(\{G_m\}_{m=k+1}^{\infty}))D_{G_k^*} \cdots D_{G_1^*} h. \end{aligned}$$

On the other hand, for $n > k+1$,

$$D_{G_k^*} H(\{G_m\}_{m=k+1}^{n})^* H(\{G_m\}_{m=k+1}^{n}) D_{G_k^*} = H(\{G_m\}_{m=k}^{n})^* H(\{G_m\}_{m=k}^{n})$$

and when $n \to \infty$ it follows that

$$D_{G_k^*} H_\infty^2(L(\{G_m\}_{m=k+1}^{\infty})) D_{G_k^*} = H_\infty^2(L(\{G_m\}_{m=k}^{\infty})).$$

By Proposition A.2, there exist the contractions r_k (actually, r_k are partial isometries) such that

$$r_k H_\infty(L\{G_m\}_{m=k}^{\infty})) = H_\infty(L(\{G_m\}_{m=k+1}^{\infty}))D_{G_k^*}.$$

Thus

$$H_\infty(L(\{G_m\}_{m=k+1}^\infty))D_{G_k^*}\cdots D_{G_1^*} = r_k r_{k-1}\cdots r_1 H_\infty(L)$$

so that

$$Z_k H_\infty(L)h = G_k H_\infty(L(\{G_m\}_{m=k+1}^\infty)) r_k r_{k-1}\cdots r_1 H_\infty(L).$$

Defining

$$D_{*,k+1} = H_\infty(L(\{G_m\}_{m=k+1}^\infty)) r_k r_{k-1}\cdots r_1,$$

we finally obtain:

Lemma A.10 For $k \geq 1$, $Z_k = G_k^* D_{*,k+1}$.

Notes

Theorem A.1 is a well—known result about positivity, as fruitful as it is simple. Proposition A.3 has the same roots (see for instance [46]). Proposition A.2 appears in [36] and different variants of Therem A.7 are discussed in many papers such as [57], [2], [41], [49], [10], [30], [69] and [9]. Lemma A.5, Proposition A.6 and Lemma A.10 are taken from [23]. The rest of the material is contained in [9].

Bibliography

[1] ADAMJAN,V. M., AROV, D.Z. and KREIN, M.G. Infinite Hankel matrices and generalized problems of Carathéodory–Fejér and I. Schur, *Funkcional. Anal. i Prilozhen* **2**(1968), 1-17 (Russian).

[2] ADAMJAN, V.M., AROV, D.Z. and KREIN, M.G. Infinite Hankel block matrices and related problems of extension, *Izv. Akad. Nauk Armjan. SSR Ser. Mat.* **6**(1971), 87–112 (Russian).

[3] AKHIEZER, N.I. *The Classical Moment Problem*, Moscow, 1961 (Russian).

[4] ANDO, T., CEAUŞESCU, Z. and FOIAŞ, C. On intertwining dilations, II, *Acta Sci. Math. (Szeged)* **39** (1977), 3–14.

[5] ANTOULAS, A.C. and BISHOP, R.H. Continued–fraction decomposition of linear systems in the state space, *System Control Lett.* **9** (1987), 43–53.

[6] AROCENA, R. Generalized Toeplitz kernels and dilations of intertwining operators, *Integral Equations Operator Theory*, **6** (1983), 759–78.

[7] AROV, D. Z. Darlington's method in the study of dissipative systems, *Dokl. Akad. Nauk SSSR* **201** (1971), 559–62 (Russian).

[8] AROV, D.Z. and KREIN, M.G. On computations of entropy functionals and their minimums, *Acta Sci. Math. (Szeged)* **45** (1983), 51–66 (Russian).

[9] ARSENE, GR., CEAUŞESCU, Z. and FOIAŞ, C. On intertwining dilations VIII, *J. Operator Theory* **4** (1980), 55-91.

[10] ARSENE, GR. and GHEONDEA, A. Completing matrix contractions, *J. Operator Theory* **7** (1982), 179-89.

[11] BAKONYI, M. Spectral factors and analytic completion, *Integral Equations Operator Theory* **13** (1990), 149–64.

[12] BALL, J.A. and HELTON, J.W. A Beurling–Lax theorem for the Lie group $U(m, n)$ which contains most classical interpolation theory, *J. Operator Theory* **9** (1983), 107–42.

[13] BAXTER, G. A convergence equivalence related to polynomials orthogonal on the unit circle, *Trans. Amer. Math. Soc.* **99** (1961), 471–87.

[14] BOYD, D.W. Schur's algorithm for bounded holomorphic functions, *Bull. London Math. Soc.* **11** (1979), 145–50.

[15] BÖTTCHER, A. and SILBERMANN, B. *Invertibility and Asymptotics of Toeplitz Matrices* Akademie Verlag, Berlin, 1983.

[16] DE BRANGES, L. Appendix on square summable power series, *Proc. Symp. Pure Math.* **Vol XI**, Amer. Math. Soc., Providence, RI, 1968.

[17] BURG, J. P. Maximal Entropy Spectral Analysis, Ph.D. Dissertation, Stanford University, 1975.

[18] CARATHÉODORY, C. Über den Variabilitätsbereich der Koeffizienten von Potenzreinen, die gegebene Werte nicht annehmen, *Math. Ann.* **64** (1907), 95–115.

[19] CARATHÉODORY, C. and FEJÉR, L. über den Zusammenhang der Extremen von harmonischen Funktionen mit ihren Koeffizienten und über den, Picard–Landauschen Satz, *Rend. Circ. Mat. Palermo* **32** (1911), 218–39.

[20] CEAUŞESCU, Z. and FOIAŞ, C. On intertwining dilations, V, *Acta Sci. Math. (Szeged)* **40** (1978), 9–32.

[21] CEAUŞESCU, Z. and FOIAŞ, C. On intertwining dilations, VI, *Rev. Roumaine Math. Pures Appl.* **23** (1978), 1471–82.

[22] CONSTANTINESCU, T. On the structure of positive Toeplitz forms, In *OT 11*, Birkhäuser Verlag, Basel, 1983.

[23] CONSTANTINESCU, T. On the structure of the Naimark dilation, *J. Operator Theory* **12** (1984), 159–75.

[24] CONSTANTINESCU, T. An algorithm for the operatorial, Carathéodory–Fejér problem, In *OT 14*, Birkhäuser Verlag, Basel, 1984.

[25] CONSTANTINESCU, T. A maximum entropy principle for contractive intertwining dilations, In *OT 23*, Birkhäuser Verlag, Basel, 1987.

[26] CONSTANTINESCU, T. On a general extrapolation problem, *Rev. Roumaine Math. Pures Appl.* **32** (1987), 509–21.

[27] CONSTANTINESCU, T. Operator Schur algorithm and associated functions, *Math. Balkanica* **2** (1988), 244–252.

[28] CONSTANTINESCU, T. Some aspects of nonstationarity, I. INCREST Preprint, No. 9 (1988); II, INCREST Preprint, No. 43 (1989), to appear In *Math. Balkanika.*

[29] COTLAR, M. and SADOSKY, C. On the Helson–Szegö theorem and a related class of modified Toeplitz kernels, *Proc. Symp. Pure Math.* **Vol XXXV**, Amer. Math. Soc., Providence, RI, 1979.

[30] DAVIS, C. KAHAN, W. M. and WEINBERGER, H. F. Norm–preserving dilations and their applications to optimal error bounds, *SIAM J. Numer. Anal.* **19** (1982), 445–69.

[31] DELSARTE, Ph., GENIN, Y. and KAMP, Y. Shur parametrization of positive definite block–Toeplitz systems, *SIAM J. Appl. Math.* **36** (1979), 34–46.

[32] DEVINATZ, A. The factorization of operator valued functions, *Ann. of Math.* **73** (1961), 458–95.

[33] DEWILDE, P. *Roomy scattering matrix synthesis*, Techn. Rept. Dept. of Maths., Univ. of California at Berkeley, 1971.

[34] DEWILDE, P., VIEIRA, A.C. and KAILATH, T. On a generalized Szegö–Levinson realization algorithm for optimal linear predictors based on a network synthesis approach, *IEEE Trans. Circuits and Systems* **25** (1978), 663–75.

[35] DEWILDE, P. and DYM, H. Lossless chain scattering matrices and optimum linear prediction: The vector case, *Circuit Theory and Applications* **9** (1981), 135–75.

[36] DOUGLAS, R.G. On majoration, factorization, and range inclusion of operators in Hilbert space, *Proc. Amer. Math. Soc.* **17** (1966), 413–15.

[37] DOUGLAS, R.G. On factoring positive operator functions, *J. Math. and Mech.* **16** (1966), 119–26.

[38] DOUGLAS, R.G. and HELTON, J.W. Inner diltions of analytic matrix functions and Darlington synthesis, *Acta Sci. Math. (Szeged)* **34** (1973), 61–7.

[39] DYM, H. J. Contractive Matrix Functions, Reproducing Kernel Hilbert Spaces and Interpolation, *CBMS Regional Conference Series* **71**, American Mathematical Society, Providence, RI, 1989.

[40] DYM, H., GOHBERG, I. Extensions of band matrices with band inverses, *Linear Algebra Appl.* **36** (1981), 1–24.

[41] DYM, H., GOHBERG, I. Unitary interpolants, factorization indices and infinite block Hankel matrices, *J. Funct. Anal.* **54** (1983), 229–89.

[42] ERDÖS, P. and TURÁN, P. On interpolation, III, *Ann. of Math.* **41** (1940), 510–53.

[43] FEDČINA, I.P. A criterion for the solvability of the Nevanlinna–Pick tangent problem, *Mat Issled.* **7** (1972), 213–27 (Russian).

[44] FRANCIS, B.A., HELTON, J.W. and ZAMES, G. H^∞– optimal feedback controllers for linear multivariable systems, In *Mathematical Theory of Networks and Systems*, Springer Verlag, Berlin, 1984.

[45] FREUD, G. *Orthogonal Polynomials*, Akademiai Kiadó/Pergamon Press, Budapest, 1971.

[46] GANTMACHER, F.R. and KREIN, M.G. *Oscillating Matrices and Kernels, and Small Oscillations of Mechanical Systems*, Moscow, 1950 (Russian).

[47] GOHBERG, I., KAASHOEK, M.A. and LERER, L. On minimality in the partial realization problem, *System Control Lett.* **9** (1987), 97–104.

[48] GERONIMUS, Ya. L. *Orthogonal Polynomials*, Consultant Bureau, New York, 1961.

[49] IANOVSKAIA, R. N. and ŠMULIAN, Iu. L. On matrices whose entries are contractions, *Izv. Vyssh. Uchebn. Zaved. Mat.* **7** (1981), 72–75 (Russian).

[50] IBRAHIMOV, I. and SOLEV, V. On the regularity of Gaussian stationary processes, *Dokl. Akad. Nauk SSSR* **185** (1971), 509–12 (Russian).

[51] HELSON, H. and LOWDENSLAGER, D. Prediction theory and Fourier series in several variables, I, II, *Acta Math.* **99** (1958), 165–202;106 (1961), 175–213.

[52] KAILATH, T. *Linear Systems*, Prentice–Hall, Englewood Cliffs, N.Y., 1980.

[53] KAILATH, T. A theorem of I. Schur and its impact on modern signal processing, In *OT 18*, Birkhäuser Verlag, Boston, 1986.

[54] KAILATH, T. and BRUCKSTEIN, A.M. Naimark dilations, state–space generators and transmission lines, In *OT 17*, Birkhäuser Verlag, Boston, 1986.

[55] KAILATH, T. and PORAT, B. State–space generators for orthogonal polynomials, In *Prediction Theory and Harmonic Analysis*, North–Holland, Amsterdam, 1983.

[56] KALMAN, R. E., FALB, P. F. and ARBIB, M. A. *Topics in Mathematical System Theory*, McGraw–Hill, New York, 1969.

[57] KREIN, M. G. The theory of self–adjoint extensions of semi–bounded Hermitian transformations and its applications, *Math. Sb* **20** (1947), 431–95 (Russian).

[58] KREIN, M. G. On Hermitian operators whose deficiency indices are 1, *Dokl. Akad. Nauk. SSSR* **43** (1944), 323–26 (Russian).

[59] KREIN, M. G. Sur le problème du prolongement des fonctions hermitiennes positives et continues, *C. R. (Dokl.) Acad. Sci. URSS* **26** (1940), 17–22.

[60] KREIN, M. G. and LANGER, H. On some continuation problems which are closely related to the theory of operators spaces Π_k, IV, *J. Operator Theory* **13** (1985), 299–417.

[61] KREIN, M. G. and NUDELMAN, A. A. *The Markov Problem of Moments and Extremal Problems*, Moscow, 1973 (Russian).

[62] KREIN, M. G. and SPITKOVSKII, I. M. On some generizations of Szegö's first limit theorem, *Analysis Math.* **9** (1983), 23–41 (Russian).

[63.] LOWDENSLAGER, D. On factoring matrix valued functions, *Ann. of Math.* **78** (1963), 450–45.

[64] MATÉ, A. NEVAI, P. and TOTIK, V. Asymptotics for the ratio of leading coefficients of orthogonal polynomials on the unit circle, *Constructive Approximation* **1** (1985), 63–9.

[65] MATÉ, A. NEVAI, P. and TOTIK, V. Asymptotics for orthogonal polynomials defined by a recurrence relation, *Constructive Approximation* **1** (1985), 231–48.

[66] NAIMARK, M. A. On self-adjoint extensions of the second kind of a symmetric operator, *Izv. Akad. Nauk SSSR* **4** (1940), 54-104 (Russian).

[67] NEHARI, Z. On bounded biliniar forms, *Ann. of Math.* **65** (1957), 153–62.

[68] PAGE, L. B. Applications of the Sz.–Nagy–Foiaş lifting theorem, *Indiana Univ. Math. J.* **20** (1970), 135–45.

[69] PARROTT, S. On a quotient norm and the Sz.–Nagy–Foiaş lifting theorem, *J. Funct. Anal.* **30** (1978), 331–58.

[70] POTAPOV, V. P. The multiplicative structure of J–Contractive matrix functions, *Trudy Moskov. Mat. Obshch.* **4** (1955), 125–236 (Russian).

[71] REDHEFFER, R. M. Inequalities for a matrix Ricatti equation, *J. Math. Mech.* **8** (1959), 349–77.

[72] RAHMANOV, E. A. On the asymptotics of the ratio of orthogonal polynomials, *Math. USSR Sbornik* **32** (1977), 199-213 (Russian).

[73] RAHMANOV, E. A. On the asymptotics of the ratio of orthogonal polynomials, II, *Math. USSR. Sbornik* **46** (1983), 105–17 (Russian).

[74] ROSENBLUM, M. and ROVNYAK, J. The factorization problem for non–negative operator valued functions, *Bull. Amer. Math. Soc.* **77** (1971), 287–318.

[75] ROSENBLUM, M. and ROVNYAK, J. *Hardy Classes and Operator Theory*, Oxford University Press, 1985.

[76] SARASON, D. Generalized interpolation in H_∞, *Trans. Amer. Math. Soc.* **127** (1967), 179–203.

[77] SARASON, D. Moment problems and operators in Hilbert spaces, In *Proc. of Symposia in Applied Mathematics* **Vol. 37** (1987), 54–70.

[78] SCHU . Über Potenzreinen, die im Innern des Einheissskreises beschränkt sind, I, II, *J. Reine Angew. Math.* **147** (1917), 205–32; **148** (1918), 151–63.

[79] SUCIU, I. and VALUŞESCU, I. Factorization theorems and prediction theory, *Rev. Roumaine Math. Pures Appl.* **23** (1978), 1393–1424.

[80] SZEGÖ, G., Beiträge zur Theorie der Toeplitzschen Formen, I, *Math.Ztschr* **6** (1920), 167–202.

[81] SZEGÖ, G. On certain hermitian forms associated with the Fourier series of a positive function, In *Festskrift Marcel Riesz* Lund, 1952, pp. 228–38.

[82] MONTROLL, E. W., POTTS, R. B. and WARD, J. C. Correlations and spontaneous magnetization of the two–dimension Ising model, *J. Math. Phys.* **4** (1963), 308–22.

[83] SZEGÖ, G. Orthogonal Polynomials, *Amer. Math. Soc., Colloquium Publ.* **No. 23**, 1939.

[84] SZ. NAGY, B. and FOIAŞ, C. it Harmonic Analysis of Operators on Hilbert Space, North–Holland, New York, 1970.

[85] TOEPLITZ, O. Über die Fourier'sche Entwicklung positiver Funktionen, *Rend. Circ. Mat. Palermo* **32** (1911), 191–92.

[86] WIENER, N. *Extrapolation, Interpolation, and Smoothing of Time Series*, Wiley, New York, 1949.

[87] WIENER, N. and AKUTOWICZ, E. J. A factorization of positive hermitian matrices, *J. Math. Mech.* **8** (1959), 111–121.

[88] WIENER, N. and MASANI, P. R. The prediction theory of multivariate stochastic processes, I, II, *Acta Math.* **98** (1957), 111–50; *99* (1958), 93–137.

[89] ZAMES, G. Feedback and optimal sensitivity: model reference transformations, multiplicative seminorms, and approximate inverses, *IEEE Trans. Aut. Control* **AC–26**, 1981, 301-20.

[90] ZASUHIN, V. N. On the theory of multidimensional stationary random processes, *Dokl. Akad. Nauk SSSR* **33** (1941), 435–37 (Russian).

[91] YOULA, D. C. JABR, H. A. and BONGIORNO, J. J. Modern Wiener–Hopf design of optimal controllers I, II, *IEEE Trans. Aut. Control* **AC–21** (1977), 3–13; (1977), 319–38.

Subject index